Dark Caves, Bright Visions

Life in Ice Age Europe

Randall White

This book is based on an exhibition presented by The American Museum of Natural History in collaboration with

Musée des Antiquités	Musée de l'Homme	Musée d'Aquitaine
Nationales	Paris	Bordeaux
Saint Germain-en-Laye		

The American Museum of Natural History
in association with
W.W. Norton & Company
New York   London

This book was supported in part by
contributions from the Richard Lounsbery
Foundation and Mr. and Mrs. Gordon P.
Getty.

ISBN: 0-393-02410-5

Library of Congress Catalog Card
Number: 86-71456

Design by Katy Homans.
Typesetting by Unicorn Graphics.
Printing by South Sea International Press.

The American Museum of Natural History
Central Park West at 79th Street
New York, New York   10024

W. W. Norton & Company, Inc.
500 Fifth Avenue, New York, NY   10110

W. W. Norton & Company Ltd.
37 Great Russell Street
London   WC1B3NU

1 2 3 4 5 6 7 8 9 0

Published simultaneously in Canada by
Penguin Books Canada Ltd., 2801 John
Street, Markham, Ontario L35 1B4.

Editor's note:
The numbers following the object descriptions
in the captions refer to the checklist.

Contents

Dedication

In addition to honoring the diligence, ingenuity, and outright brilliance of late Ice Age humans, the present volume is a tribute to these same traits shown by prehistorians and anthropologists. The very general synthesis provided here relies heavily—often without the detailed credits seen in more technical works—on generations of work by researchers who devoted their lives to the subject: Absolon, Breuil, Peyrony, Semenov, Capitan, Bordes, Laming, Vertut, Begouen, Lartet, and Piette are only a few of the great investigators who are no longer with us.

However, during the past 30 years or so, one person stands out above all others in contributing to technical and methodological advances in our understanding of late Ice Age prehistory. Our paths never crossed. I knew him only through his work, which I have relied upon heavily in writing this volume. Each rereading of his *Préhistoire de l'art occidental* provides new insights and overlooked details. His ideas have become part of the consciousness of most Paleolithic archeologists. In always seeking the reconstruction of human behavior, he avoided many of the typological entanglements of traditional French prehistory. This book and the exhibition that it accompanies salute Professor André Leroi-Gourhan, one of the great anthropologists of this century. His work lives on.

R.W.

Acknowledgments

No exhibition as complex as this one could have been achieved without the enthusiasm and effort of many people. *Dark Caves, Bright Visions: Life in Ice Age Europe* was initially inspired by a previous show, *Art et Civilisations des Chasseurs de la Préhistoire,* mounted at the Musée de l'Homme, Paris, by Professor Henry de Lumley and his colleagues. The successful implementation of the New York exhibition, though different in concept from its predecessor, owes much to the active . assistance of Professor de Lumley, Mme. Marie Perpère, and many others at the Musée de l'Homme.

That we have been able to base *Dark Caves, Bright Visions* almost entirely on the display of original specimens is due in great part to the vision and support of representatives of the Musée des Antiquités Nationales in Saint Germain-en-Laye, which possesses the world's most extensive collection of Upper Paleolithic portable art. We especially thank M. Henri Delporte, Chief Curator of the Museum, and his colleague, M. Jean-Jacques Cleyet-Merle, for their generosity and spirit of international collaboration.

For other noteworthy loans from France, we are indebted to M. Alain Roussot, Curator of Prehistory at the Musée d'Aquitaine in Bordeaux, whose support has allowed us to exhibit the Venuses from Laussel. We also wish to thank M. Michel Soubeyran of the Musée du Périgord, Périgueux; M. Jacques Allain of Argenton-sur-Creuse; Mme. Marie-Françoise Poiret of the Musée de Brou, Bourg-en-Bresse; Mme. Beatrice Schmider and Mlle. Yvette Taborin of the Université de Paris I; Dr. Jean Gaussen of Neuvic-sur-Isle; Mme. Arlette Leroi-Gourhan of the Musée de l'Homme, Paris; Professor Philippe Taquet of the Institut de Paléontologie, Muséum d'Histoire Naturelle, Paris; and M. Jean-Philippe Rigaud, Directeur des Antiquités Préhistoriques d'Aquitaine, Bordeaux.

Other European lenders include Professor Jan Jelinek of the Moravske Museum, Brno; Professor Hans Müller-Beck and Dr. Joachim Hahn of the Universitat Tubingen; Professor Fausto Zevi of the Museo Preistorico e Etnografico, Rome; Professor Paolo Graziosi,

Instituto di Antropologia, Firenze; and Dr. L. Sorbini, Museo Civico di Storia Naturale, Verona.

Cooperating institutions in North America include the Logan Museum of Anthropology, Beloit College; the Field Museum of Natural History, Chicago; the Royal Ontario Museum, Toronto; and the Peabody Museum of Archaeology and Ethnology, Cambridge.

The successful realization of *Dark Caves, Bright Visions* owes much to the enthusiastic support of Robert G. Goelet, President of the American Museum, and Dr. Thomas D. Nicholson, Director. Many others at this museum have worked tirelessly to make the exhibiiton a success, none more than Paul Beelitz, Assistant Registrar, who cheerfully shouldered an enormous burden of paperwork and logistics. The success of the efforts of Ralph Appelbaum and his design team from Ralph Appelbaum Associates will be evident to everyone who visits *Dark Caves, Bright Visions.* Lyn Hughes of New York University provided the title of the exhibition. The exhibition is supported by an indemnity from the Federal Council on the Arts and Humanities.

Specifically with regard to this publication, many contributions must be recognized. The photographs and drawings were generously provided by a number of scholars and technicians, whose names appear in the figure captions. John Pfeiffer made helpful comments on the text. Logistic help in acquiring photographs was provided by Mme. Yvonne Vertut, Mme. Marie Perpère, M. Denis Vialou, and M. Alain Roussot. At Natural History Magazine, Tom Kelly, Mark Abraham and Sherry Krukwer Sundel and Colleen Mehegan patiently brought this book to fruition. Margaret Cooper and Nancy Creshkoff did an admirable job of turning academic jargon into English. The design of the volume is the work of Katy Homans and her assistant Mark La Rivière. Production of the catalogue was made possible by the generosity of the Richard Lounsbery Foundation and of Mr. and Mrs. Gordon P. Getty.

We take pleasure in expressing our gratitude to all of these individuals and institutions and regret that space precludes mentionir the names of many others who have provided invaluable assistance.

Randall White, Guest Curator

Ian Tattersall, Organizing Curator

Foreword

A major presentation of Paleolithic art in the United States in 1986!
Such an ambitious project undoubtedly responds to one of the deepest
desires of many American archeologists. It was also our desire, when
confronted with the possibility in the summer of 1985—so long as the
most prestigious of western European, and especially French, Upper
Paleolithic art objects could be presented to American researchers and
public alike in the way that these objects deserve. The exhibition had
to be spectacular and scientifically rigorous at the same time. This has
been accomplished, thanks to the initiative taken by the American
Museum of Natural History in New York, in collaboration with the
Musée des Antiquités Nationales in Saint Germain-en-Laye and the
Musée de l'Homme in Paris, as well as numerous other museums and
individuals on both sides of the Atlantic.

The extraordinary record of Franco-Cantabrian *cave art*—of
which Lascaux is probably the best-known example—is known in
every country on earth. In contrast, late Upper Paleolithic *portable
art*—which provides primary evidence of Magdalenian culture as the
first of the world's great civilizations—is widely unappreciated. The
diversity and splendor of portable art, exemplified in the pieces
chosen for this exhibition, are well known only to specialists, despite
the art's importance as an essential and complementary aspect of
Magdalenian artistic achievement. The Musée des Antiquités
Nationales is the principal repository of archeological collections of
portable art from southwestern France, the cradle of world prehistoric
science. Beginning in the late nineteenth century, renowned
prehistorians such as E. Lartet, E. Piette, E. Massenat, P. Girod, R. and
S. Saint-Perier, M. and St. Just Pequart, G. and H. Martin, S. de Saint-
Mathurin, R. Robert, and many others assembled the museum's
remarkable holdings. The museum houses nearly half of the several
thousand known Paleolithic art objects, including most of the greatest
works from this period. As a result, it plays a crucial role in the
American Museum's undertaking.

A special effort was required of the Musée des Antiquités Nationales to allow its most remarkable pieces to come to New York to be presented for the first time in the United States. Because of the extreme fragility of these pieces, they are stored under rigorous environmental conditions—inaccessible to visitors and seen in replica in the galleries at Saint Germain-en-Laye.

A spirit of close collaboration has emerged between the large Paris museums (Musée des Antiquités Nationales, Musée de l'Homme) and those in the rest of the country (Musée d'Aquitaine, Musée du Périgord, and others). Despite all of the imaginable difficulties, none hesitated to make the sacrifice of temporarily giving up crucial parts of its collection. The result is evident and requires no further comment.

The interest of this international exhibition considerably exceeds the simple presentation of a unique assemblage of prehistoric art objects. Well before preparations began in France, R. White traversed North America and uncovered in museum storerooms objects whose abundance was not previously recognized. One might say that he "rediscovered" many major works that had been inadequately published at the beginning of the century or that were totally unknown in Europe. As a result, limestone engravings from Limeuil and Abri Blanchard, as well as one of the female statuettes from Grimaldi—all first-class pieces housed in North American institutions—will occupy choice locations beside their French counterparts in the American Museum of Natural History. (Author's note: The Grimaldi Venus and a number of other objects uncovered by Marshack (1981) appeared in the American Museum of Natural History *Ice Age Art* exhibition in 1978.)

Such is the step that we take together toward spreading knowledge about the hunting cultures of the European Paleolithic. This step opens up avenues for productive collaboration between New and Old World researchers. Joint study of previously dispersed archeological materials will provide immense mutual enrichment; for example, the majority of engravings from the Abri Blanchard are in France, but much of the bone and lithic material is at Beloit College, the Field Museum, and the American Museum of Natural History. The same is true of the fragmentary engraved pieces from Limeuil; some forty undeciphered engravings are housed at the Logan Museum of Anthropology at Beloit College, while the majority of Limeuil plaques are being studied at the Musée des Antiquités Nationales. The international effort launched by the Centre d'Information et Documentation (C.I.D. Breuil) to inventory, describe, and computerize French, German, Spanish, and other Paleolithic portable art objects will find in the United States a vast field of inquiry and will remedy the dispersal of collections from classic sites in the Périgord.

Thus, the exhibition at the American Museum of Natural History is the materialization of a common wish. In conclusion and in the name of the Direction des Musées de France and the French public, we want to express another wish—that the art works and accompanying archeological material held by American museums will temporarily abandon their adoptive country to return to, for a few months and in a great exhibition, their place of origin.

H. Delporte
Inspector General of Museums

J.-J. Cleyet-Merle
Curator

Musée des Antiquités Nationales
Saint Germain-en-Laye

Preface

Every human society is fascinated by its own origins and has created elaborate myths to explain how humanity, so close to and yet so distinct from other living organisms, came to be. As we near the close of the 20th century, we are fortunate to be able to derive our explanations of how humanity evolved, both physically and culturally, from a rich and ever-increasing store of paleontological and archeological evidence. Presenting and interpreting this evidence to the public has been among the chief preoccupations of the American Museum of Natural History during its almost 120 years of existence, and during 1984 we were extremely proud to present to the public our ground-breaking exhibition *Ancestors: Four Million Years of Humanity*. In that exhibition, for the first time anywhere, a substantial proportion of the best-known and most complete of the fossils that document the story of human biological evolution were displayed together, allowing the public to appreciate for itself the quality of the evidence upon which our understanding of human evolution is based.

Today, in the same spirit, we are equally honored to present *Dark Caves, Bright Visions: Life in Ice Age Europe*. This exhibition also documents, through the display of original materials, the emergence in Europe of fully modern humans. With the birth of our own kind came what has been called a "creative explosion"—an outburst of inventiveness and sensibility entirely without previous parallel in the long history of the human family. Evidence of this flowering is known from many regions of the world, but nowhere is it more comprehensive and impressive than in Europe. It is the ready cooperation of our European, notably French, sister institutions (individually identified and acknowledged elsewhere in this book) that has made this exhibition possible. We are also, of course, most grateful to those museums on this side of the Atlantic that have generously loaned European materials in their care to us. In thus bringing together objects that document the same cultural developments but that are normally separated by an ocean, *Dark Caves, Bright Visions* also performs an important scientific service in allowing them to be appreciated and compared side by side.

Dark Caves, Bright Visions, then, permits us to contemplate in the original some of the finest creations of our early progenitors, works that rival both in imagination and in execution anything that has been achieved since. Daily life is not forgotten, however, and we can equally discern the roots of the common humanity that unites us all, everywhere, in the humbler productions of these early people.

Robert G. Goelet
President of the Board of Trustees
The American Museum of Natural History

CHAPTER 1

The Upper Paleolithic: A Human Revolution

Fig. 1 The original site of Cro-Magnon in Les Ey-zies-de-Tayac, France. In 1868, under the rock overhang hidden by the modern hotel, workers building the railroad in the foreground uncovered the remains of four adults and a very young infant or fetus dated to the Aurignacian period, about 30,000 years ago. At a time of running debate over Darwinian evolution and how ancient the earliest humans might be, these remains provided important evidence of the association of physically modern humans with long extinct animals and "primitive" stone technology.

Introduction

This book celebrates the accomplishments of anatomically and culturally modern humans as they existed in Europe from 35,000 to 12,000 years ago, a period known to archeologists as the Upper Paleolithic, or late Ice Age. The record we have found of their life provides one of the most exciting stories of new ideas and practical advances in the whole span of the human past. During this time, we see a virtual explosion of symbolic behavior, best exemplified by the cave and portable art of western Europe. This revolutionary period of human achievement contrasts sharply with the longer, quite stagnant period beginning around 100,000 years ago or earlier, when the Neanderthals lived scattered over many of the same areas.

The term "Ice Age," known technically as the Pleistocene, refers to a period of marked climatic cooling that began some time before 1.5 million years ago. Within this period, there were many fluctuations between warmer and cooler conditions. The cooling trend was more intense during the past 500,000 years, when massive ice sheets descended from the polar region at intervals, only to retreat during subsequent warmer periods. The last major glacial advance, known in Europe as the Würm glaciation, began about 75,000 years ago and ended about 12,000 years ago. It is the last half of this advance that interests us here and that we will refer to as "the late Ice Age."

The term "modern humans" describes us and ancestors anatomically like ourselves, members of the sub-species *Homo sapiens sapiens*. We have inherited the physical characteristics and cultural capacities of these ancestors. They have passed down to us abilities and capacities that we modern humans take for granted. These include the capability to speak and comprehend symbolically based language, the capacity to monitor and adjust behavior according to a framework of shared norms and values, the ability to imagine things that have never been observed, and the ability to externalize physical skills in the form of tools. As the anthropologist Leslie White once put it, in understanding cultural differences among modern humans, biology can be considered a constant.

Not so when we examine the full sweep of hominid evolution over the past 2 million years. Our earliest tool-making ancestors of 2 million years ago were very different from us in stature, in manual dexterity, and in brain volume. Indeed, marked differences persisted until relatively recent times. Even a cursory examination of a 100,000-year-old Neanderthal skeleton alongside that of a modern human provides a striking contrast (Fig. 2,3). Given such clear physical differences, when we compare the material patterns left behind by our much more distant ancestors, we must consider the possibility that they did not share all of our capabilities. A few chipped stones at Olduvai Gorge in levels dated to nearly 2 million years ago cannot be taken to imply culture as it is exhibited by modern humans and as it was defined by Edward Burnett Tylor (1871):

. . . that complex whole which includes knowledge, belief, art, morals, custom and any other capabilities and habits acquired by man as a member of society.

If culture as defined by Tylor is characteristic of humans, we are confronted by a problem. As we proceed back in time, we encounter an archeological record that suggests the absence of some of Tylor's cultural attributes and other characteristics that we assume to be exhibited by all "human" societies. According to what framework can we understand prehistoric populations that produced no observable technological inventions over thousands of generations?

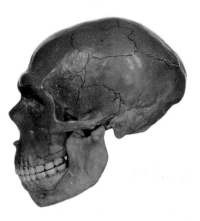

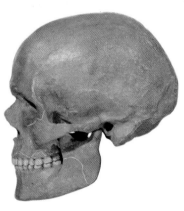

Figs. 2,3 Skulls of a Neanderthal and a modern human showing major differences in morphology. Courtesy Department of Library Services A M N H

or that did not exchange or trade goods with other such groups? or hunt but instead lived by scavenging the carcasses of already dead animals? or decorate themselves? or that did not produce artistic images in material media? Indeed people prior to 35,000 years ago (the boundary may well be much earlier in the Near East and Africa) were so different from us—both biologically and culturally—that we may well ask whether it is appropriate to apply the term "human" to them. Such observations are the basis for an evolutionary perspective regarding human culture.

Neanderthal Ancestors

By at least 100,000 years ago, a physical type known as Neanderthals (*Homo sapiens neanderthalensis*) occupied Europe (Trinkaus and Howells, 1979; Stringer et al., 1984; Smith, 1984). In general, Neanderthals were considerably more robust than we. They were much more muscular, especially in the neck and shoulder region, than all but the most massive of modern humans. The walls of their bones were very thick. Partly because of their overall massiveness, Neanderthals had greater cranial volumes than ours. One of their most obvious traits was a bony ridge above the eyes, which, combined with a protruding lower face, gave them their very distinctive appearance.

Recently, a provocative hypothesis has been put forth by Erik Trinkaus of the University of New Mexico (Trinkaus, 1984). On the basis of his careful examination and measurement of pelvises and skulls of numerous Neanderthal infants, he suggests that the Neanderthal gestation period may have been much longer than ours. He proposes a period of 12 or 13 months. If true, this would mean much greater prenatal development and a shorter postnatal maturation period. This evidence is a striking reminder of how different the Neanderthals may have been from us. However, this remains a rather controversial hypothesis, not yet widely accepted in paleoanthropology.

Another intriguing observation by Trinkaus is that Neanderthals used their teeth as tools, producing an odd wearing down of their

Figs. 4–7 Examples of stone tools made by Neanderthals, all from the very rich Mousterian site of Combe-Capelle Bas (Dordogne), France. Typically they are made on rather large, chunky "flakes" of flint. All loaned by Royal Ontario Museum. Photos by R. White.

Fig. 4 Scraper made from flint flake. 8.9 cm long. (223)

Fig. 5 Hand axe, a kind of tool formed by re-moving flakes from both faces. 12.3 cm long. (200)

incisors, but not in the way seen among modern Eskimos, who use their teeth to soften hides by chewing and for a variety of other tasks. The precise activity that produced this wear remains to be explained, but it implies to some anthropologists that Neanderthals solved fewer problems with technology than do their descendants.

The material record left by Neanderthals between 100,000 and 35,000 years ago is known as the Mousterian period (Bordes, 1968, 1972). The record for the Mousterian contrasts sharply with that of the modern humans who followed. While the Neanderthals made a variety of tools in stone—some of them quite complicated—the lack of innovation over more than 50,000 years is surprising from a modern perspective. For example, not a single tool form existed 40,000 years ago that was not already present 60,000 years earlier. In the words of the great French prehistorian François Bordes, "They made beautiful tools stupidly." He meant that there was an almost

Fig. 6 Flake cleaver formed by removing a series of small flakes from both faces of one edge, resulting in a durable, sharp, jagged cutting or chopping edge. 6.7 cm long. (225)

Fig. 7 Large, thick scraper on the side of a flake, formed by removing a number of small flakes on one face, thus providing a strong, steep angle suitable for a variety of scraping tasks. 11.9 x 7.0 cm. (202)

mechanical redundancy to Mousterian tools that may imply more programmed behavior than that of later people. Mousterian stone technology is generally described as a "flake technology." This means simply that the final, carefully worked tools are chipped out of rather amorphous rough outs, or blanks, about as long as they are wide (Fig 4–7). These are removed from large globular chunks of stone known as "flake cores."

Also surprising is the similarity of Mousterian stone tools from region to region. In form and relative numbers, they are often virtually identical across vast distances; collections from sites in France often cannot be distinguished from those found in the Near East, for example (Bordes, 1968).

From the beginning of the Upper Paleolithic in Europe, modern humans used a wide variety of materials to manufacture tools and other objects. The Mousterians did not. Few examples of objects in bone, antler, and ivory have been discovered in Mousterian sites; even these are hotly disputed by specialists. Wood—if it was used as a raw material—would not have endured over the years.

Finally, evidence for any kind of activity that we would call artistic is almost nonexistent in Mousterian sites. Fewer than a dozen Mousterian objects bear even the simplest markings. We do begin to find evidence in this period for the collection of natural pigments in the form of chunks of ocher and manganese, but the context of their use remains uncertain. Some show evidence of having been applied to soft surfaces such as human or animal skin, but this practice does not necessarily imply artistic activity. We now know that ocher very effectively repels vermin and prevents animal skins from decaying. If Neanderthals were using it for these purposes, then they seem not to have practiced any form of bodily ornamentation—unlike all modern humans.

Lewis Binford of the University of New Mexico observes that Neanderthal use of fire was very simple (Binford, 1982). Hearths were small and usually not surrounded by heat-retaining stones (Perlès, 1977). Binford also suggests that actual hunting by Neanderthals and all of their ancestors may not have been very extensive. Many bones in Mousterian sites indicate that other predators had killed and partially consumed the animals. There are instances where cut marks from stone tools overlay tooth marks from predators on fragments of animal bone, indicating that the humans got to the animal after the predator had done so. Also, the body parts left at living sites are those that scavengers usually retrieve from carcasses and that have already been partially consumed. Not all paleoanthropologists agree. The evidence is still sketchy enough to support various views, such as the hunting of large game animals from central base camps, as proposed by the late Glyn Isaac (1978).

Few constructed shelters are known for the Mousterian. This absence may result from a lack of preservation, since large numbers of Mousterian sites exist in exposed areas, suggesting that Neanderthals did not need caves or rock shelters to survive. The few examples of shelters that exist were made from durable materials, such as animal bone. It may simply be that shelters were made of wood or other nonpreservable materials. Indeed, there is evidence much earlier for wooden stakes and post molds (de Lumley, 1969; Bordes, 1961).

Neanderthals made tools from stone that almost always came from the immediate vicinity of their campsites. This pattern has led Binford to argue that their tool manufacture and use was expedient; that is, they did not make long-term plans but carried tools as they moved about and discarded them immediately after use. This idea of limited planning is perhaps supported by the Neanderthals' apparent failure to profit by cyclical patterning in the environment. Migratory fish, which provide a rich resource at the same time each year, were not exploited, for example. Nor do Neanderthals seem to have taken advantage of the highly predictable migrations of reindeer, which became very important for modern humans.

If certain universal characteristics of modern humans were lacking among Mousterians, others were clearly present. Notably, it is in the Mousterian (but probably only after about 60,000 to 50,000 years ago) that we find the first purposeful and ritualized burial of the dead. Graves frequently include animal parts, often interpreted by archeologists as offerings for use in life after death (Harrold, 1980). In addition, this burial activity indicates a conventionalized set of ways for dealing with death—a disruptive event in all human societies. The very fact of establishing a convention suggests a cultural coding and externalizing of emotional responses and a shared body of ideas.

The Mousterian period came to an end around 35,000 years ago. After 60,000 or more years of a stable technology and way of life, the end was sudden—at least by evolutionary standards. Specialists have argued back and forth for decades: Was the abrupt disappearance of the Mousterians the result of an invasion of peoples from elsewhere, or was it a matter of *in situ* evolution? Until recently, most archeologists agreed that the Neanderthals were responsible for the Mousterian, and that the succeeding cultures were the product of anatomically modern humans. However, this sequence is no longer certain. At the site of Saint-Césaire (ApSimon, 1980; Lévêque and Vandermeersch, 1982) in southwestern France, remains of two

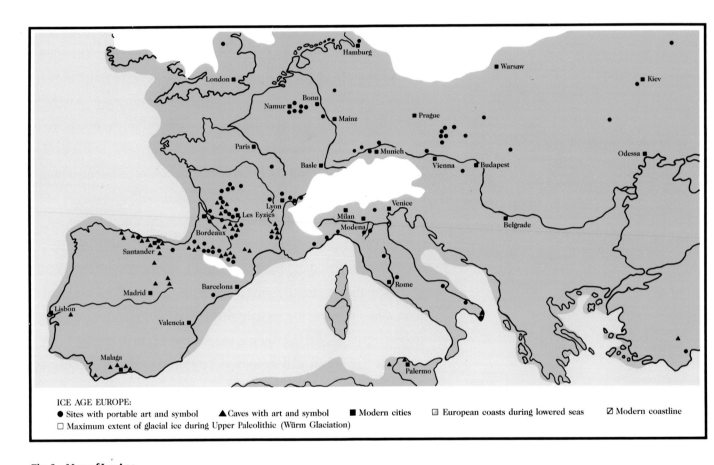

ICE AGE EUROPE:
● Sites with portable art and symbol ▲ Caves with art and symbol ■ Modern cities ▢ European coasts during lowered seas ▨ Modern coastline
▢ Maximum extent of glacial ice during Upper Paleolithic (Würm Glaciation)

Fig. 8 Map of Ice Age Europe showing locations of important sites and position of ice front during maximum glaciation. © Alexander Marshack, 1972 from *The Roots of Civilization*.

Neanderthals were recently excavated. Archeologists were in for a big surprise. The skeletons were in a level containing Upper Paleolithic tools and dated to about 32,000 years ago. This unexpected find suggests that the swift transition to the Upper Paleolithic may not have been the result of intrusive populations. However, all known skeletal specimens after 32,000 years ago are those of *Homo sapiens sapiens*.

Such complexity in Europe should have been expected. From other regions, such as Africa, the Near East and central and eastern Europe, it is becoming clear that the temporal boundary between archaic forms of *Homo sapiens* (such as Neanderthals) and fully modern humans is not the same everywhere. Both anatomically modern humans and Upper Paleolithic cultural assemblages occur in eastern Europe and the Near East significantly earlier than in western Europe. Considerable evidence is emerging from Africa and even Australia for anatomically modern humans prior to those of the European Upper Paleolithic. The richness of the western European record has led to a bias in favor of only one set of regional developments.

However, Europe is the focus of our discussion here. Therefore, we can sidestep some of these complex issues. *It is nevertheless important to emphasize that Europe had no monopoly on important evolutionary trends. Exciting evidence, both cultural and biological, is being sought and found in virtually every part of the globe.*

The Environment

The Upper Paleolithic odyssey must begin with an understanding of what Europe was like before, during, and after the emergence of modern humans. When we think of European climate, vegetation, and landscape today, the word "diversity" clearly comes to mind. The same has been true for the past 100,000 years, and the conditions in any given region have changed again and again (Gerasimov and Velichko, 1982).

Huge ice sheets, today restricted to the polar areas, periodically expanded to cover areas farther south. Most of northern Europe was covered by a mile-high sheet of ice during periods of maximal glacial

Fig. 9 Galloping reindeer (*Rangifer tarandus*) engraved on a slab of schist. 12.1 x 8.6 x 1.7 cm. Saint-Marcel (Indre), France. Late Magdalenian. One of the most remarkable and lifelike art objects of the entire Upper Paleolithic. Note the execution of the antlers and hoofs and the lightly pecked surface suggesting the appearance of the fur. A natural spall detached during the engraving process forced the artist to foreshorten the lower jaw. Musée des Antiquités Nationales. Photo by MAN. (73)

advance (Fig. 8). There were several periods during the long span of the Upper Paleolithic when the ice receded to nearly its present range. During glacial retreat, more northerly areas were available for habitation by prehistoric people, but subsequent readvances of ice destroyed most traces of these early occupations.

So much of the earth's water was locked up in glacial ice during the periods of advance that world sea levels dropped by as much as 100 meters (300 vertical feet). As a result, both the English Channel and the Bering Sea were dry land and were occupied by humans for long periods. When the glaciers finally retreated about 12,000 years ago, world sea levels rose again and covered all evidence of human occupation in these areas with fathoms of water.

South of present-day Paris and London, traces of early peoples' activities are well preserved since the glaciers never descended this far. In these areas, modern studies of fossil pollen, geological sediments, and animal bones have yielded a clear picture of continual environmental change. When the glaciers reached their maximum, climate in southern Europe was cold and dry. Broad expanses of steppe (rich grasslands), always with at least some trees, covered much of unglaciated Europe. At no time during the past 100,000 years was widespread arctic tundra a dominant feature of Europe.

Fig. 10   Shaggy male horse (*Equus caballus*) painted in black and red pigment. Approximately 70 cm long. Niaux Cave (Ariège), France. Middle Magdalenian. The rock contours have been used to suggest that the animal is standing on the ground surface. A "wound" appears behind the shoulder. Photo by J. Vertut.

Local topography has long made the environmental picture very complicated. In areas of high relief, vegetation has varied considerably between river-valley bottoms and uplands, for example. In southwestern France, even during the maximum glacial advance, trees requiring mild conditions survived in warmer microclimates such as south-facing slopes. In general, this rugged countryside exhibited much diversity in plant communities.

Varied vegetation in a region allows a wide variety of animals to coexist (Figs. 9–17). During much of the late Ice Age, in hilly to mountainous terrain like that of Cantabrian Spain, southwestern France, southern Germany, and Czechoslovakia, large herbivores (as well as carnivores and scavengers) with very different environmental requirements coexisted (Freeman, 1973; Straus, 1983). These included, for example, mountain goats, reindeer, bison, and horses

Fig. 11 Bison (*Bison priscus*). **About 2 m long. Painted in polychrome on cave ceiling. Altamira Cave (Santander), Spain. Middle Magdalenian. The outlines have been carefully drawn in black and then filled in with red ocher, differentially applied to convey shading and coat characteristics. This bison forms part of an immense composition that includes numerous bison and other animals in lifelike postures. Photo by J. Vertut.**

(Delpech, 1983). The prehistoric human occupants shared these landscapes with as many as a dozen species of such large-bodied animals, a situation unknown in these latitudes today. Far from being impoverished, these environments had a higher animal biomass than any occupied by hunting and gathering peoples today. It is more reasonable to compare these environments with the modern plains of Africa than with present-day arctic tundras.

In regions with more even topography, diversity of species was reduced, but the animal biomass remained very high (Klein, 1973; Gladkih, Kornietz, and Soffer, 1984; Soffer, 1985). Less varied environmental conditions meant fewer kinds of animals but more of each kind. The gigantic woolly mammoths seem to have been few in number in the rugged countryside of southwestern France but

Fig. 12   Running male aurochs (*Bos primigenius*) and two superimposed cows painted in black on cave wall. About 4.2 m long. Lascaux Cave (Dordogne), France. Lower Magdalenian. The horns show what Breuil (1952) described as twisted perspective. A forked sign appears in front of the bull. Natural coloration was used for the ground surface. Photo by J. Vertut.

abundant on the broad, flat steppes of eastern Europe, for example.

During times when the great masses of ice shrank back to the polar regions, European environments changed accordingly. The climate was much less dry and generally warmer, with temperatures even reaching today's levels during the warmest periods, when much of Europe became forested. Animals such as horses and mammoths, which had previously thrived on dry grasslands, could not cope with the new forests, and their numbers were diminished. These animals were replaced by such animals as red deer (*Cervus elaphus*) and wild boar (*Sus scrofa*), which flourish in wet, heavily forested conditions (Delpech et al., 1983).

Because animal bones are well preserved for thousands of years and plant remains are not, archeologists tend to underestimate the

Fig. 13 Woolly mammoth (*Mammuthus primigenius*) engraved on cave wall, showing shaggy coat in great detail. About 69 cm long. Rouffignac Cave (Dordogne), France. Magdalenian. Mammoths differed from modern elephants in having a domelike head and a large humped back. Engraved signs can be seen in this photograph. Lumpy protrusions are large nodules of flint. Photo by J. Vertut.

potential contribution of plant and other foods to the Ice Age diet. In fact, the pollen grains recovered from late Ice Age sites in Europe include those of a number of edible species of tubers, nuts, and berries that were growing in the vicinity. These include such items as blueberries, raspberries, acorns, and hazelnuts. Some of the Ice Age art clearly indicates an interest in plants, not to mention amphibians and insects, probably due to their importance as foodstuffs. In recent excavations at the 15,000-year-old site of El Juyo, in northern Spain, L. Freeman and J.G. Echegaray have recovered a number of seeds (Fig. 21) through use of a modern technique known as flotation. Included are the seeds of blackberry, raspberry, and grasses.

Fig. 14 Ibex (*Capra ibex*) painted in black on cave wall. About 49 cm long. Niaux Cave (Ariège), France. Middle Magdalenian. Ibexes are adapted to rugged, rocky terrain, and their bones are often found in the deposits of Pyrenean Upper Paleolithic sites. Photo by J. Vertut.

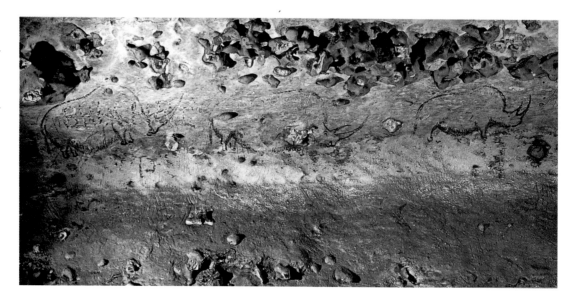

Fig. 15 Three woolly rhinoceroses (*Coelodonta antiquitatis*) in procession, painted in black on cave wall. Each about 1 m long. Rouffignac Cave (Dordogne), France. Magdalenian. These animals, rarely encountered in the bone refuse of Upper Paleolithic sites, must have been a very dangerous quarry. Photo by J. Vertut.

27

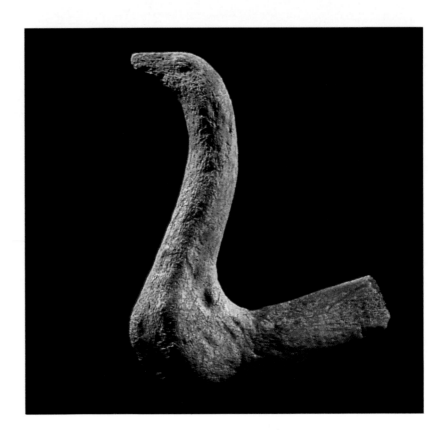

Fig. 16 Lion (*Panthera spelaea* or *Felis sp.*) engraved on shaft of broken pierced baton fragment. About 20 cm long. Laugerie-Basse (Dordogne), France. Middle to Late Magdalenian. The only animal of this type known to have adorned a pierced baton, it resembles the North American cougar or mountain lion except for the bobbed tail. Musée de l'Homme. Photo by J. Oster, MH. (159)

Fig. 17 Long-necked bird sculpted from bone or antler. 4.5 cm long. Laugerie-Basse (Dordogne), France. Magdalenian. While some birds are found in Upper Paleolithic sites of all periods, they become much more abundant during the Magdalenian. Musée de l'Homme. Photo by J. Oster, MH. (172)

The picture of the past 100,000 years in Europe that emerges, then, is one of near-continuous climatic oscillation, sometimes rendering parts of northern Europe uninhabitable. In any given region, there were several periods over that time span during which humans would have been required to readapt to changing environmental conditions by altering their diet, their hunting and gathering strategies, their technology, and their knowledge of the world around them.

The Upper Paleolithic Revolution

If the Mousterian saw a glimmer of change toward what we recognize as human behavior, the Upper Paleolithic saw an explosion—an explosion that continued for about 250 centuries until the end of the

Fig. 18　Grasshopper and partial birds engraved on bone, one of the few Upper Paleolithic artworks showing an insect. 10 cm long. Enlène Cave (Ariège), France. Magdalenian. According to Leroi-Gourhan, "At top left we see the wing and belly of a bird, probably a sparrow, beak open to catch a grasshopper. At top right are the legs and lower body of another bird. Turning the object around, we find a third bird at top right: it has the beak of a sparrow, pecking at small round things, perhaps seeds. Its legs are not visible, and under its belly are two little bag-shaped or larva-shaped bodies resembling ant eggs. Lastly, at top left, a fourth bird shows the rear of its body and one leg, possibly a sparrow's." Musée de l'Homme. Photo by J. Oster, MH. (185)

Fig. 19 Fragment of thick bone splinter, probably once part of a spear-thrower. 5.6 cm long. Laugerie-Basse (Dordogne), France. Middle to Late Magdalenian. It is decorated with several images of fish, one engraved and at least two sculpted. Fish, especially seasonally migratory varieties, became a very important food source toward the end of the Upper Paleolithic. Musée de l'Homme. Photo by J. Oster. (166)

Fig. 20 Salamander sculpted on reindeer antler object of which it forms one end. The back, belly, and sides are covered with punctuations representing details of the animal's skin. Small amphibians and reptiles are frequently found in the bone debris of Paleolithic living sites, but it is uncertain how important they were to the diet. 15 cm long. Laugerie-Basse (Dordogne), France. Middle to late Magdalenian. Musée de l'Homme. Photo by J. Oster, MH. (162)

Ice Age created a very different set of environmental conditions for human adaptation. The pattern that begins 35,000 years ago in Europe is one of frequent—if not continual—change in human behavior (White, 1982). The succeeding 25,000 years are subdivided into cultural periods (Movius, 1973), each with its own style of technology and each characterized by a set of innovations (Fig. 23). Many of these new ways of doing things were undoubtedly stimulated by practical need. But, in many respects, the evidence gives the impression of change for the sake of change—something familiar to most modern humans as "fashion."

The generally recognized cultural periods for the Upper Paleolithic are as follows. The dates, in years before present (BP), are approximate and vary from region to region.

Chatelperronian	*35,000–30,000**
Aurignacian	*34,000–30,000**
Gravettian	*30,000–22,000*
Solutrean	*22,000–18,000*
Magdalenian	*18,000–11,000*
Azilian	*11,000–9,000*

* In western Europe, these two different types of artifact assemblage coexist during the early part of the Upper Paleolithic. The significance of this overlap is not yet fully understood.

Fig. 21 Seeds, including sticktights, blackberry/raspberry, grasses, and dock. El Juyo (Santander), Spain. Magdalenian. Recent refinement of flotation procedures by Leslie Freeman and Jesus Echegaray in their excavations has led to recovery of seeds such as these and other macrobotanical remains.

Such research is helping to overcome the long-standing impression of Upper Paleolithic people as predominantly hunters rather than hunter-gatherers. Photo courtesy of L. Freeman. © Institute for Prehistoric Investigations.

Along with frequent change through time, major regional differences emerged. Different regions of Europe each had their own peculiar cultural characteristics at any given time, so much so that each region's stone technology has to be described with a different classification scheme. The Solutrean, for example, is characterized by carefully worked bifacial points and does not exist outside of France and Spain. In central and eastern Europe, local variants of the Gravettian continue throughout the period occupied by the Solutrean in the west. In Italy, the period contemporary with the Late Magdalenian elsewhere is known as the Romanellian. In central Europe, the Gravettian shows several regional variants (Otte, 1981). It is as if barriers to communication were arising, perhaps as the result of the development of regional dialects.

Archeological sites dating to the period after 35,000 years ago provide a radically different material record from that of the Mousterian. Part of what is new and revolutionary is that, unlike the Mousterian, the Upper Paleolithic can be understood by using the same analytical and interpretive vocabulary that anthropologists apply to human cultures of the 20th century.

Fig. 22 The relationship between climatic changes and major cultural periods of the Upper Paleolithic in Europe.

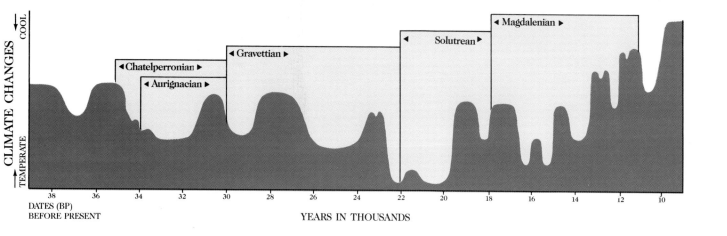

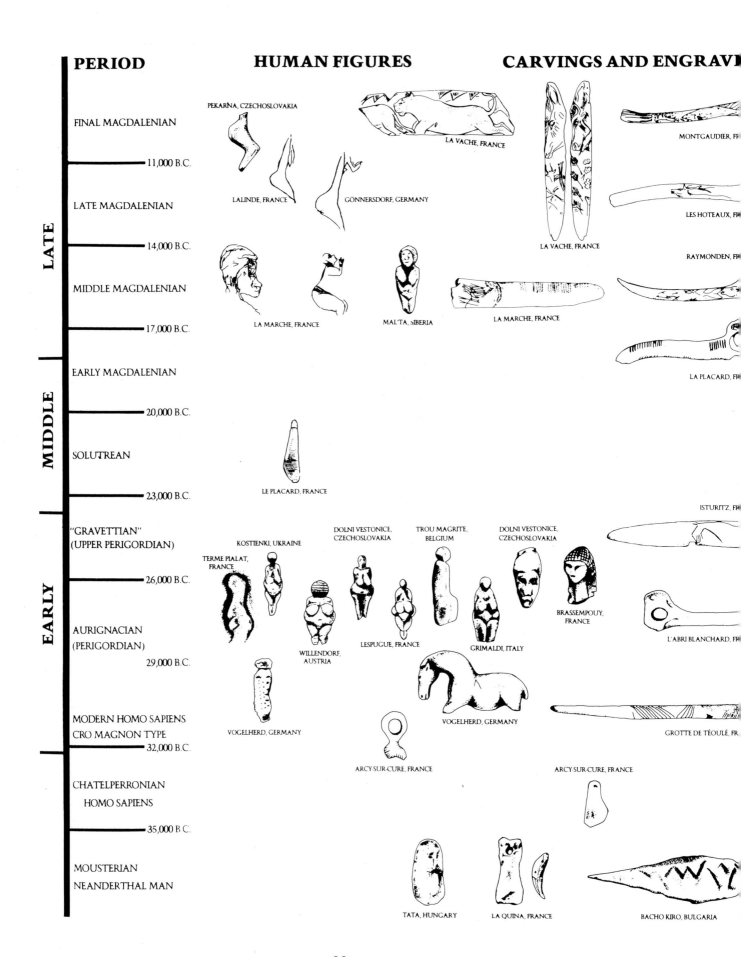

PERIOD

HUMAN FIGURES

CARVINGS AND ENGRAVI

LATE

FINAL MAGDALENIAN

— 11,000 B.C.

LATE MAGDALENIAN

— 14,000 B.C.

MIDDLE MAGDALENIAN

— 17,000 B.C.

MIDDLE

EARLY MAGDALENIAN

— 20,000 B.C.

SOLUTREAN

— 23,000 B.C.

EARLY

"GRAVETTIAN"
(UPPER PERIGORDIAN)

— 26,000 B.C.

AURIGNACIAN
(PERIGORDIAN)

29,000 B.C.

MODERN HOMO SAPIENS
CRO MAGNON TYPE

— 32,000 B.C.

CHATELPERRONIAN
HOMO SAPIENS

— 35,000 B.C.

MOUSTERIAN
NEANDERTHAL MAN

PEKARNA, CZECHOSLOVAKIA

LA VACHE, FRANCE

MONTGAUDIER, FR

LALINDE, FRANCE

GÖNNERSDORF, GERMANY

LES HOTEAUX, FR

LA VACHE, FRANCE

RAYMONDEN, FR

LA MARCHE, FRANCE

MALTA, SIBERIA

LA MARCHE, FRANCE

LA PLACARD, FR

ISTURITZ, FR

DOLNI VESTONICE,
CZECHOSLOVAKIA

TROU MAGRITE,
BELGIUM

DOLNI VESTONICE,
CZECHOSLOVAKIA

KOSTIENKI, UKRAINE

TERME PIALAT,
FRANCE

BRASSEMPOUY,
FRANCE

L'ABRI BLANCHARD, FR

WILLENDORF,
AUSTRIA

LESPUGUE, FRANCE

GRIMALDI, ITALY

VOGELHERD, GERMANY

VOGELHERD, GERMANY

GROTTE DE TÉOULÉ, FR

VOGELHERD, GERMANY

ARCY-SUR-CURE, FRANCE

ARCY-SUR-CURE, FRANCE

LE PLACARD, FRANCE

TATA, HUNGARY

LA QUINA, FRANCE

BACHO KIRO, BULGARIA

SIGNS—SYMBOLS NOTATIONS

CUETO DE LA MINA, SPAIN

MARSOULAS, FRANCE

EL CASTILLO, SPAIN

TROIS FRÈRES, FRANCE

FONT-DE-GAUME, FRANCE

LASCAUX, FRANCE

EL CASTILLO, SPAIN

LA PILETA, SPAIN

LAUGERIE HAUTE, FRANCE

LA FERRASSIE, FRANCE

GRIMALDI, ITALY

ABRI BLANCHARD, FRANCE

LA FERRASSIE, FRANCE

WALL ART

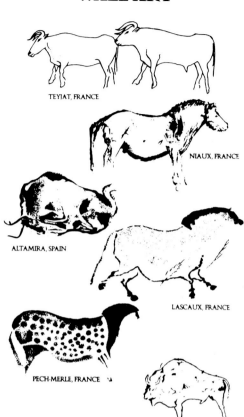

TEYJAT, FRANCE

NIAUX, FRANCE

ALTAMIRA, SPAIN

LASCAUX, FRANCE

PECH-MERLE, FRANCE

LA GRÈZE, FRANCE

GARGAS, FRANCE   PECH-MERLE, FRANCE

GARGAS, FRANCE

Fig. 23 The chronology of Upper Paleolithic cultures and their related art forms. © Alexander Marshack, from the *Ice Age Art* exhibition, AMNH 1978.

CHAPTER 2

Strategies for Survival

Fig. 24 Part of living surface at Etiolles (Essonne) France (about 13,000 years old). This site has yielded 20 buried living surfaces with fireplaces, piles of chipping debris from stone tool manufacture, and abundant tools and animal bones. Near the fireplace covered by a large slab of rock (bottom center), a zone interpreted as a work area shows a dense concentration of stone tool chipping debris and ocher. Beyond, an accumulation of stone blocks may have supported a post, perhaps part of a dwelling structure. In the background is a dumping ground for stone debris removed from living and work area. Photo and description courtesy of Y. Taborin.

Introduction

Our most detailed and reliable knowledge about life in the Upper Paleolithic concerns the ways in which people made a living; the story is one of creativity and innovation. That story is as impressive as their artistic accomplishments. Indeed, it will be evident as we proceed that art and survival were tightly entwined. Tools and weapons were often much more elaborate than their function required and often carried designs and images of no technological value that indicate a complex body of ideas and conceptions. What is centrally evident is that the Upper Paleolithic represents a technological revolution after hundreds of thousands of years that saw little or no change in technology or subsistence strategies.

Tool-Making and Use

The first thing apparent about Upper Paleolithic technology is that tool-makers no longer produced stone tools mostly on amorphous flakes but on blades. Blades are much longer than they are wide, and some of them are as thin and as sharp as modern knives. While such blades were sometimes made in the Mousterian, only later were they produced systematically and continuously.

In general, the number of stages between the conception of a tool and its completion was much greater in the Upper Paleolithic than in the Mousterian. This process included the meticulous and time-

35

consuming preparation of the stone core from which blades can be struck. Blades were then transformed into the desired tool form by "retouching"—chipping away at the edges with a stone, antler, or wooden hammer. By the end of the Magdalenian, skill in blade production was remarkable. At the site of Etiolles, near Paris (Fig. 24) blades as long as 60 centimeters (24 inches) have been recovered (Pigeot, 1983).

The ability to produce thin, scalpellike blades rather than thick flakes meant a much more efficient use of raw material. By some measures, Upper Paleolithic craftspeople could obtain ten times more usable "cutting edge" than could Mousterian tool-makers from the same amount of flint. This advantage would have been especially important in areas where stone was scarce or during winter, when stone sources would have been covered by snow.

Archeologists have divided the stone tools found at Upper Paleolithic sites into more than 100 different types. However, these types fall into a very few more general categories. The most frequentl encountered of these are "scrapers," "burins," and "perforators" (Figs. 25, 26). It is unfortunate that these categories were named before we knew anything about the actual uses for the tools. On the basis of microscopic examination of wear and polish on tool edges, w now know that the purpose suggested to us by the shape of a tool was not necessarily the one that the tool actually served. Many examples are now known of burins (meaning "chisel" or "engraving tool") used to scrape bone or hide, scrapers used to gouge or shave a variety of materials, and perforators used as chisels (Keeley, 1980).

Beginning in the Solutrean, a very sophisticated technique of stone preparation was developed. Known as heat treatment, this process involved "cooking" large pieces of flint for fairly long periods The end result is a raw material with a porcelainlike sheen from which flakes can be removed quite easily. Indeed, the development of heat treatment in the Solutrean seems to have coincided with a mean of working flint, known as "pressure flaking," by which the tool-maker forced flakes off a piece of flint by exerting pressure with a "flaker" made from a hard substance such as reindeer antler or hardwood. Many of the fine Solutrean leaf-shaped points were produced in this fashion (Figs. 29, 30).

Fig. 25 Typical examples of scrapers on the end of flint blades. Left: 10.5 cm long. Abri du Roc-Tombé (Dordogne), France. Aurignacian. This scraper shows scaly retouch along the edges, a characteristic of the Aurignacian period. A white patina covers the flint, which has changed from its originally blue/black color. Right: 9.7 cm long, provenience unknown, probably Magdalenian. This specimen is of a different variety of flint. Royal Ontario Museum. Photos by R. White. (211, 226)

Fig. 26 Flint with "burin" formed at top and scraper at bottom. 7.5 cm long. Le Moustier (Dordogne), France. Solutrean. Such multiple tools are common in the Upper Paleolithic. Used as incising tools, burins also served for many other tasks, including production of art. Bouyssou's excavation. Royal Ontario Museum. Photo by R. White. (197)

Figs. 27,28 Two examples of flint perforators used for various drilling and perforating tasks in working bone, antler, wood, and animal skin. 27: The more substantial type of perforator. 11 cm long. La Madeleine (Dordogne), France. Magdalenian. 28: Delicately worked perforator on a fine, translucent brown flint blade. 9.1 cm long. Le Soucy (Dordogne), France. Late Magdalenian. Both from Royal Ontario Museum. Photos by R. White. (220, 213)

26

27

28

Figs. 29,30 Characteristic examples of Solutrean flint working. 29: Flint point carefully worked in light-colored flint, probably by pressure flaking. 9.3 cm long. Roc de Combe-Capelle (Dordogne), France. Late Solutrean. The point was broken off in prehistoric times. 30: Small leaf-shaped flint point made from gray/brown flint. 5.8 cm long. Plateau de la Sellerie (Dordogne), France. Solutrean. As with the other example, this point was carefully retouched on both faces. Royal Ontario Museum. Photos by R. White. (216, 215)

Fig. 31 Flint bladelet still embedded in soil cast of what was probably a wooden handle. About 3.5 cm long. Lascaux Cave (Dordogne), France. Early Magdalenian. Many of the flint tools from Lascaux had preserved on their surfaces some form of adhesive (probably pitch or sap) for fixing the tool to its handle. The remarkable preservation at this site has contributed much to understanding Early Magdalenian stone technology. Musée de l'Homme. Photo by J. Oster, MH. (148)

29

30

31

Figs. 32a,b   Front and side views of "navette," of reindeer antler, probably a tool handle. About 20 cm long. Roc de Marcamps (Gironde), France. Middle Magdalenian (phase III). A stone scraper or burin, for example, would have been squeezed into the fork at either end and bound tightly with sinew or cord. Musée d'Aquitaine. Drawing by A. Roussot. (141)

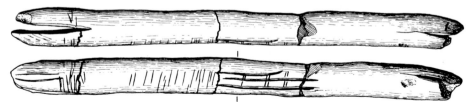

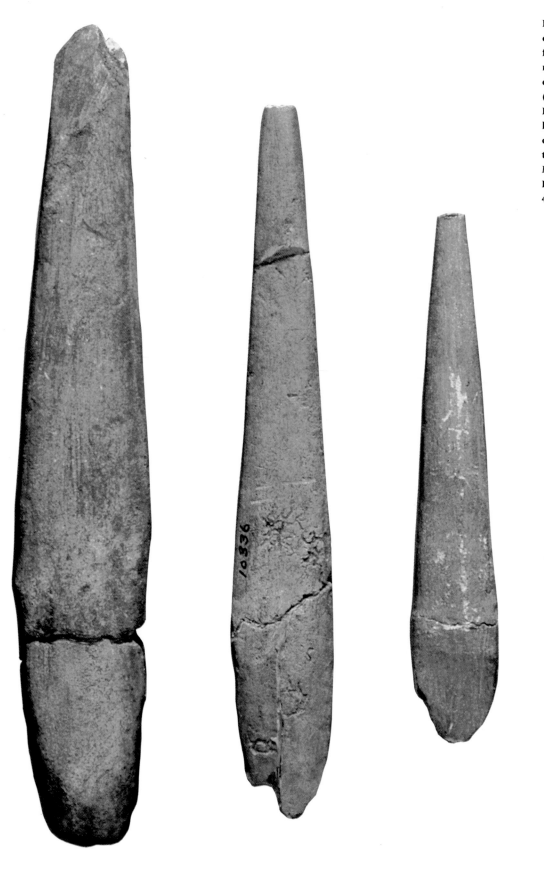

Figs. 33-36 Spear-points of varying sizes made from ribs, long bones, or reindeer antler. 7–15.8 cm long. Abri Cellier (Dordogne), France. Early Aurignacian. These have a split base to accommodate the end of the spear shaft. Logan Museum, Beloit College. Photo by R. White. (42, 43, 44, 41)

Figs. 37–40 Engraved spear-points of bone or antler. 9.7 to 14.0 cm long. From left: (37) Rivière de Tulle (Lot), France, (38) and (39) Crozo de Gentillo (Lot), France. All Magdalenian. (40) Jouclas (Lot), France. Solutrean. Examples 38-40 are fragments. Their bases have a two-sided bevel to be fitted into the slitted end of a spear shaft. The marks on (37) probably helped keep the point fixed in the spear shaft. Engraved marks on the others perhaps served as identification for individuals or social groups. All loaned by Logan Museum, Beloit College. Photos by R. White. (51, 53, 61, 58)

Fig. 41 Sculpted bone baton known as "the scepter." 25 cm long, from deepest level in the "Salle Monique," La Vache (Arièges), France. Final Magdalenian (radiocarbon date of about 10,900 BC). Between the bird (grouse or ptarmigan?) carved at the distal end and the fish (probably of the salmon family) at the other is the profile of a deer in relief. On the opposite side, not visible here, two partial animals face away from the bird, the forequarters of a felid and a horse head with a long mane. This piece is of great interest because five different animals are shown on the same object. It shares with a baton from Teyjat (Dordogne) the association of a bird, a quadruped, and a fish. The mammal/fish theme is common in Magdalenian art of the Pyrenees and Périgord. Musée des Antiquités Nationales. Photo by MAN. Description by J. J. Cleyet-Merle. (108)

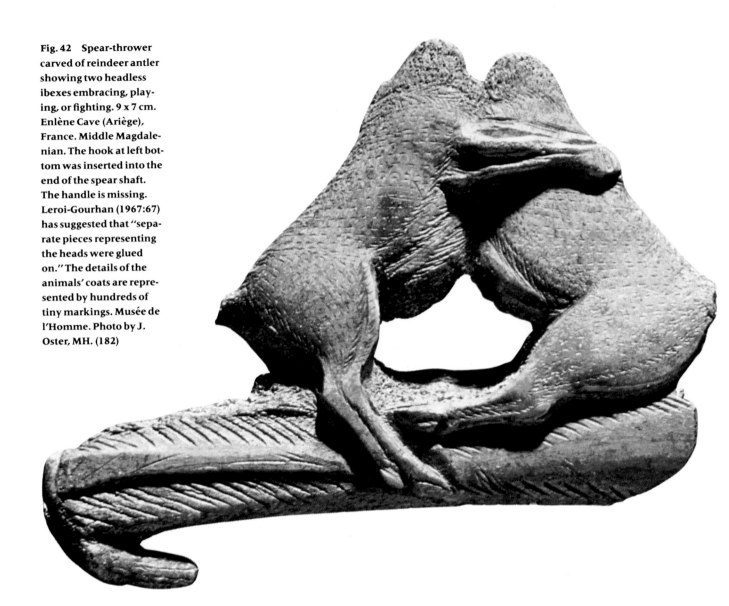

Fig. 42 Spear-thrower carved of reindeer antler showing two headless ibexes embracing, playing, or fighting. 9 x 7 cm. Enlène Cave (Ariège), France. Middle Magdalenian. The hook at left bottom was inserted into the end of the spear shaft. The handle is missing. Leroi-Gourhan (1967:67) has suggested that "separate pieces representing the heads were glued on." The details of the animals' coats are represented by hundreds of tiny markings. Musée de l'Homme. Photo by J. Oster, MH. (182)

Fig. 43 Spear-thrower of reindeer antler with carved leaping horse. The handle has been perforated. The front legs are folded against the chest. The rear legs are extended. Anatomical detail and sense of motion are remarkable. 20.7 cm long. Bruniquel (Tarn-et-Garonne), France. Middle Magdalenian. Musée des Antiquités Nationales. Photo by MAN. Description by H. Delporte. (105)

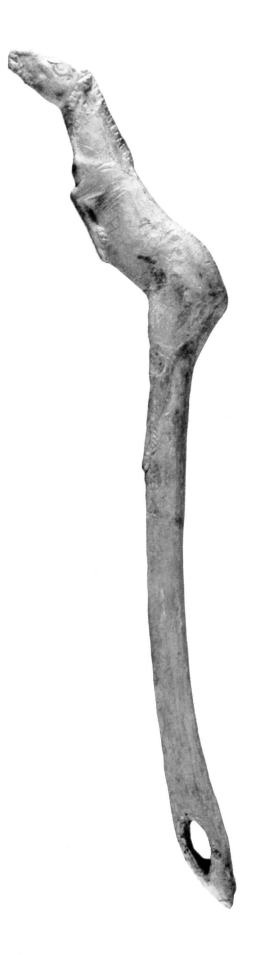

Our understanding of Upper Paleolithic stone technology remains imperfect, partly because the retouched stone blades were probably not complete tools. There is now quite early evidence for the process of hafting, in which stone tools form only one segment of a more complex tool or weapon. Frequently, the stone tools that we recover were probably attached to a handle or socket made from organic material (Fig. 32). Some 17,500-year-old flint bladelets discovered at Lascaux Cave (Leroi-Gourhan and Allain, 1979) in southwestern France still had some sort of adhesive (sap or pitch) clinging to them. In one case, what had been a wooden handle had been replaced by a perfect clay cast on the back of a bladelet (Fig. 31). Thus, the stone tools that we find have forms which may have been suitable for a number of tasks, depending upon how they were mounted in their hafts.

An important development in the Upper Paleolithic is exemplified by a series of stone artifact types inferred to have been spear-points (Figs. 33–40). The preceding Mousterian provides few convincing examples of spear-points, raising the question as to how widespread hunting was before 35,000 years ago. However, animals can be brought down using drive-and-trap techniques that do not rely on spear-points. Nevertheless, one of the most distinctive inventions of the period around 35,000 years ago was a diverse bone/antler projectile technology. Some of the earliest bone and antler spear-points have a split or forked base for inserting the end of the spear. Most of the subsequent spear-point types worked on just the opposite principle: the spear-point was made to fit into a fork or onto an oblique surface at the end of the spear shaft.

By around 20,000 years ago, Upper Paleolithic people had invented a deadly device for launching their spears with greater accuracy and velocity. This device, the spear-thrower, was a foot or two long with a handle on one end and a hook on the other that fitted into the blunt end of the spear (Figs. 42–46). This arrangement extended the throwing arc of the hunter. By the middle of the Magdalenian period, around 15,000 years ago, spear-throwers were elaborately decorated with animal forms. The same themes—for example, headless animals—often recurred.

Fig. 44   Broken spear-
thrower of reindeer ant-
ler in form of headless
ibex. 10.7 cm long. Saint-
Michel-d'Arudy (Basses-
Pyrénées), France. Mid-
dle Magdalenian.
Compare with Fig.
42.The angle of the ani-
mal's legs results from
the convergence in the
beam of a reindeer ant-
ler. The tail is raised and,
as with several other Pyrenean spear-throwers,
the animal is defecating.
Often, although less clear
in this case, a bird
perches on the emerging
excrement, its tail acting
as the hook of the spear-
thrower. Musée des
Antiquités Nationales.
Photo by MAN. (86)

Fig. 45 Broken spear-thrower of reindeer antler in form of headless ibex or deer. 8.7 cm long, Enlène Cave (Ariège), France. Middle Magdalenian. It is very similar in conception to the object in Fig. 26. Musée de l'Homme. Photo by J. Oster, MH. (183)

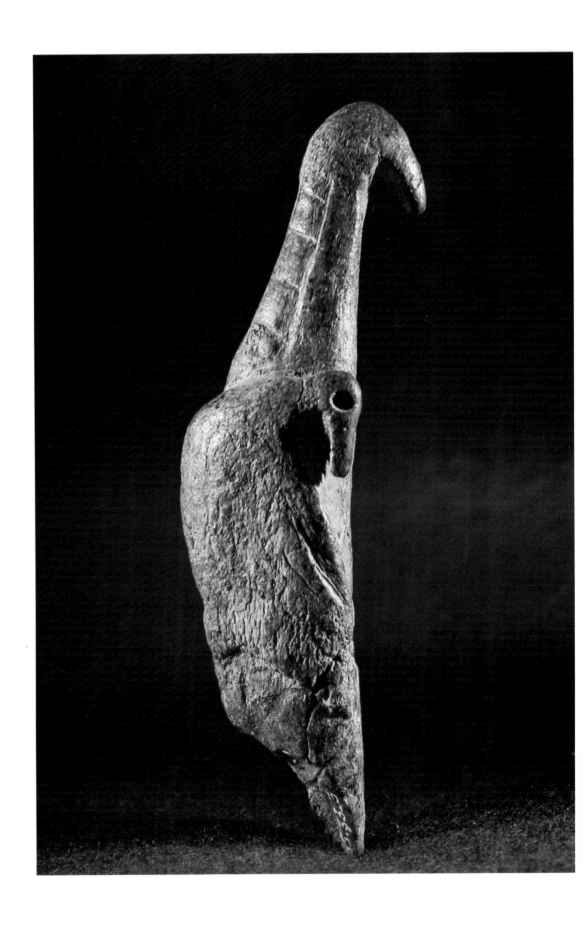

Fig. 46 Spear-thrower sculpted of reindeer antler in form of bird. 8.9 cm long. Enlène Cave (Ariège), France. Middle Magdalenian. Also visible is another bird carved in bas-relief whose head contains a hole. Another substance such as ivory may have been inserted there to form the eye, a technique often seen on spear-throwers of this period. Musée de l'Homme. Photo by J. Oster, MH. (184)

Clearly, by the end of the Upper Paleolithic, a deadly hunting technology had been developed. Sometimes we find graphic evidence for it. At the Late Magdalenian site of Reignac in the Vézère Valley of southwestern France, Alain Roussot reported (1973) the discovery of a reindeer skull in which an antler spear-point was still imbedded.

Many of the stone tools examined by archeologists were used to cut, scrape, and chop plant materials such as wood. Since plant materials are seldom preserved, many items of technology may well have existed that did not survive for archeologists to discover. But sometimes we get lucky. Shortly after the discovery of the painted cave of Lascaux in 1940, excavations at the bottom of a natural shaft in the limestone floor of the cave produced the earliest-known fragment of rope, or cord (Glory, 1959). Made of three braided plant fibers, it had been transformed to humus, and its remains were preserved in a lump of clay (Fig. 48). It is often suggested that this rope was used by early Magdalenian people to descend to the bottom of the shaft, where they painted a hunting scene (Fig. 47). But the implications of this rope go much further. Cordage must have been technologically important and probably served a variety of tasks. People who know how to make cord can easily produce nets and snares, which increase hunting efficiency enormously. Of course, nets can also be effectively employed in fishing, which became very important after about 15,000 years ago. After this time, there were also abundant bone items interpreted as fishhooks and barbed harpoons (Figs. 49–54), some of which were probably used for fishing (Julien, 1982).

The functions of many of the objects fabricated in bone, antler, and ivory remain unknown to us (Figs. 55–70). Perhaps the most puzzling of these is the pierced reindeer antler "baton," first seen at the beginning of the Upper Paleolithic and continuously present until the end. It consists of part of the shaft of a reindeer antler, perforated by drilling and often decorated with engraved or sculpted animals. Hypotheses about the purpose of these objects range from identification as symbols of leadership to spear-shaft straighteners, to handles for slings, to cheek pieces for bridle bits, to tent pegs. The answer remains elusive.

Fig. 47 The scene painted in the 10-foot-deep "well" in Lascaux Cave, (Dordogne), France. Early Magdalenian. The most literal interpretation is that the bison, disemboweled by the spear, has gored the hunter. However, a number of images and symbols that remain unexplained strongly suggest other layers of meaning: the birdlike quality of the human and his erect penis; the bird-headed stick; the barbed image; and the departing rhinoceros to the left with tail raised and six dots behind its anus. The remote and inaccessible location of the painting also seems significant. The scene probably represents a well-known myth, as a similar scene exists in the cave of Villars (Dordogne), France. Photo by J. Vertut.

Fig. 48 Lump of clay bearing the impression of a three-strand braided rope made from plant fibers. Clay 8.2 cm long. Lascaux Cave (Dordogne), France. Early Magdalenian. The rope, transformed to humus, is contained in the other half of this lump. Discovered when the clay was broken apart during excavation, it is the earliest known cordage. Its existence suggests a more complex technology than that reflected in the durable materials usually recovered. Found in the *puits,* or well, in Lascaux Cave, the rope may have been used to descend the 10 feet to the bottom of this shaft. Institut de Paléontologie Humaine. Photo by J. Oster, MH. (149)

Figs. 49–53 Tiny barbed implements or projectiles of bird bone. (49) is 4.8 cm long. (49–51) are of a type commonly found in Late Magdalenian sites with abundant fish remains, suggesting that they may have been fishhooks. (52) may be an unfinished example. (53) is a complex barbed implement of unknown function. All from Rocher de la Peine (Dordogne), France. Late Magdalenian. All loaned by Logan Museum, Beloit College. Photo by R. White. (68, 69)

Fig. 54 Harpoon of rein-
deer antler barbed on
two sides. 14.3 cm long.
Le Soucy (Dordogne),
France. Late Magdale-
nian. Such harpoons
were designed both to
penetrate an animal's
hide and to remain in
place, causing increased
bleeding. Ranging
greatly in size and form,
they may have been used
for several purposes. This
specimen, originally
much longer, suffered a
broken base; two barbs
on the right side and one
on the left were removed
to form the new base.
Royal Ontario Museum.
Photo by R. White. (221)

Fig. 55 Large fragment of pierced baton of reindeer antler, engraved with a row of four horses on the side visible here, and three on the other. 31.1 cm long. La Madeleine (Dordogne), France. Late Magdalenian. According to Delporte (1969), the antler was pierced after the horses were engraved, obliterating one of the horses. These batons, of uncertain function, are frequently decorated, especially with horses. The base of the antler (far right) originally attached to the reindeer skull. Musée des Antiquités Nationales. Photo by MAN. (102)

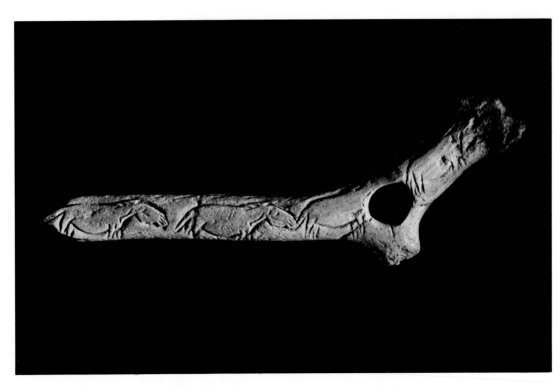

Fig. 56 Pierced baton of reindeer antler engraved on both sides and tinted with red ocher. It is broken at the perforation. 15.7 cm long. La Madeleine (Dordogne), France. Middle to Late Magdalenian. This side (right to left) shows the front of a horse's head, a simplified human figure with a "stick" over the shoulder, a second horse head, and a snake with a barbed tail surrounded by a series of deep incisions. The opposite side bears indistinct engravings of two bovids. Musée des Antiquités Nationales. Photo by MAN. (103)

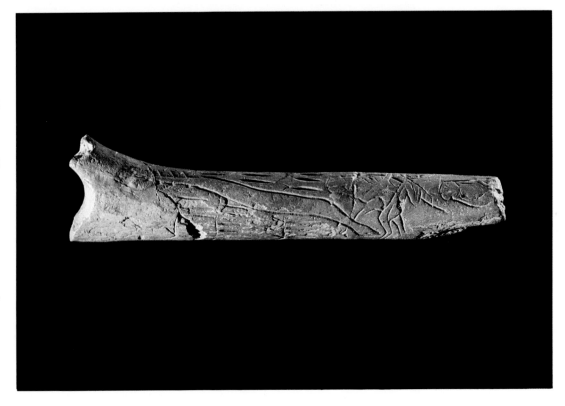

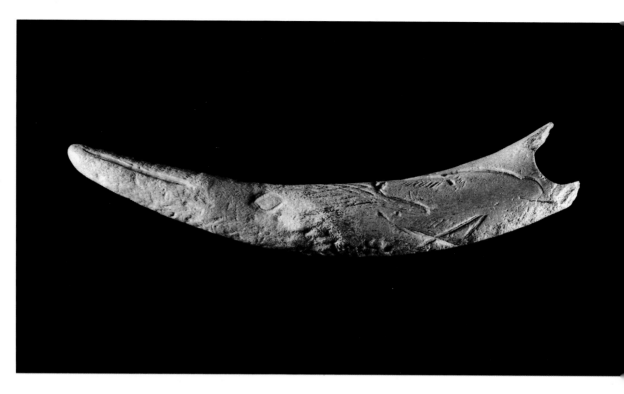

Fig. 57 Engraved pierced baton of reindeer antler, broken at the perforation. 14 cm long. Massat (Ariège), France. Magdalenian. According to Marshack's excellent analysis (1972:238–239), a bear, perhaps with blood spurting from its mouth and nose, is associated with the schematic image of a water bird. Marshack regards this as a seasonal image, and also notes an apparent barb or wound in the bear's throat. The opposite side shows the right eye and other side of bird's bill. Musée des Antiquités Nationales. Photo by MAN. (71)

Fig. 58 Broken end of a pierced baton of reindeer antler, sculpted with two bison, back to back. 12.8 cm. Laugerie-Basse (Dordogne), France. Middle to Late Magdalenian. The smaller and less detailed animal may be a female, the larger a male. This dual bison theme is common in Magdalenian art. Musée des Antiquités Nationales. Photo by MAN. (80)

Fig. 59 Small, broken pierced baton of reindeer antler, engraved on both sides. 10.6 cm long. Le Soucy (Dordogne), France. Late Magdalenian. This side shows the head and neck of a horse with the chest and upper leg extending to the edge. The neck is marked with two short parallel lines, a common feature of Magdalenian engraved horses (compare with Fig. 101). The opposite side has a roughly sketched headless body of an unidentified animal. Royal Ontario Museum. Photo by R. White. (219)

Fig. 60 Sculpted and engraved baton of reindeer antler with two perforations, known as "La chasse à l'auroch." 30.2 cm long. La Vache (Ariège), France. Late Magdalenian (radiocarbon date of 10,435 BC). One perforation, open on one end, is bordered by a sculpted cervid head in frontal view, its antlers formed by the open edges. At center is an aurochs, with massive horns presented in twisted perspective, an unexpectedly archaic technique for this period. It is followed by three hunters, also in relief, one apparently holding a bow (see detail on facing page). This scene resembles the bison hunting theme on a pendant or churinga, from Raymonden (see Fig. 119). Musée des Antiquités Nationales. Photo by MAN. Description by J.-J. Cleyet-Merle. (110)

Fig. 61 Pierced baton of reindeer antler engraved and sculpted. 23 cm long. Laugerie-Basse (Dordogne), France. Middle to Late Magdalenian. The animal theme represented has been variously described as two schematic bison, back to back, and as a crested bird with an open beak, similar to some Pyrenean examples. A continuous incision spirals around the center 11 times, much like the threads on a screw—possibly phalliform. Musée de l'Homme. Photo by J. Oster, MH. (157)

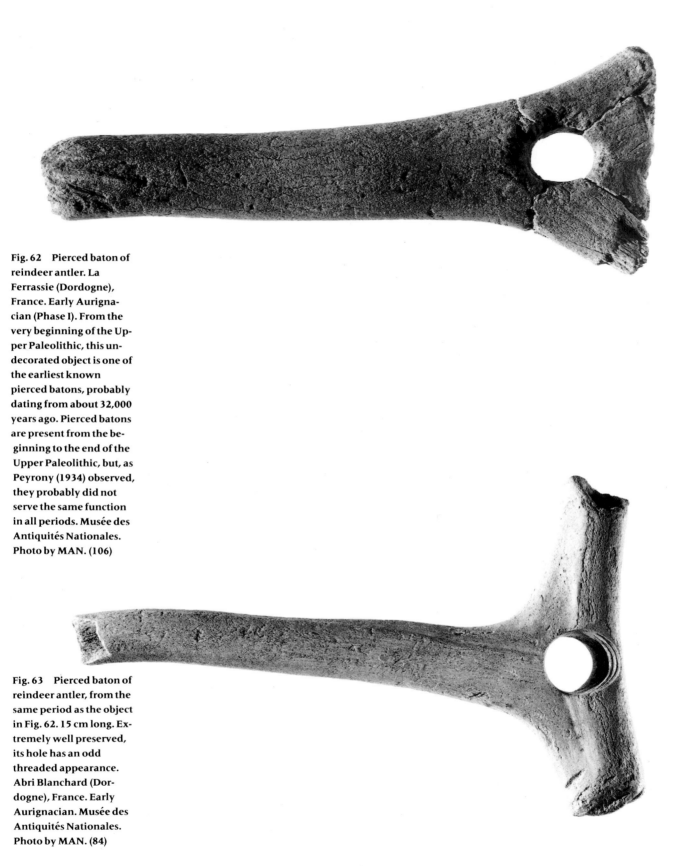

Fig. 62 Pierced baton of reindeer antler. La Ferrassie (Dordogne), France. Early Aurignacian (Phase I). From the very beginning of the Upper Paleolithic, this undecorated object is one of the earliest known pierced batons, probably dating from about 32,000 years ago. Pierced batons are present from the beginning to the end of the Upper Paleolithic, but, as Peyrony (1934) observed, they probably did not serve the same function in all periods. Musée des Antiquités Nationales. Photo by MAN. (106)

Fig. 63 Pierced baton of reindeer antler, from the same period as the object in Fig. 62. 15 cm long. Extremely well preserved, its hole has an odd threaded appearance. Abri Blanchard (Dordogne), France. Early Aurignacian. Musée des Antiquités Nationales. Photo by MAN. (84)

Figs. 64a,b Two sides of a pierced baton of reindeer antler sculpted with head of unidentifiable animal. 14.4 cm long. Le Placard (Charente), France. Solutrean. Musée des Antiquités Nationales. Photo by MAN. (82)

Fig. 65 Pierced baton of reindeer antler, engraved with baying stag having summer/autumn antlers. 24.5 cm long. Les Hoteaux (Ain), France. Late Magdalenian. In this sensitive image, the treatment of the mouth and thick neck fur are especially notable. (See Marshack 1972). Musée de Brou. Photo by MB. (142)

Fig. 66 Fragment of pierced baton of reindeer antler, sculpted with a bison head in bas-relief. 15.5 cm long; bison head about 2.5 cm high. Isturitz (Basses-Pyrénées), France. Middle Magdalenian. A masterpiece of Ice Age art, it is a fine example of the detail shown in Middle Magdalenian art of the Pyrenees. Compare with Figs. 10, 44, 45, 46. Musée des Antiquités Nationales. Photo by MAN. (113)

Fig. 67 Broken pierced baton shaft of reindeer antler with horse head engraved in relief. 16.3 cm long. Saint-Michel-d'Arudy (Basses-Pyrénées), France. Late Magdalenian. Two distinct lines above the head give the confusing appearance of antlers or horns. Another line contiguous with the horse's muzzle extends off the broken end of the piece. The eye and muzzle show remarkable detail. Musée des Antiquités Nationales. Photo by MAN. (87)

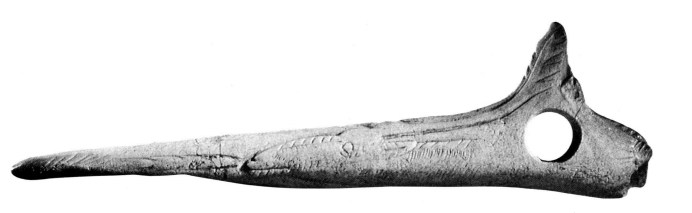

Fig. 68 Pierced baton of reindeer antler, engraved with reindeer (?) head and other images. 27 cm long. Laugerie-Basse (Dordogne), France. Middle to Late Magdalenian. The antlers seem to curve along the antler tine located beside the hole. A series of abstract signs appears in front of the snoutlike muzzle. A carved figure at extreme right may be a bison head. The reverse side bears a magnificently engraved reindeer with head turned. Musée de l'Homme. Photo by J. Oster, MH. (171)

Figs. 69,70 Two bone objects of uncertain function. (69) has a spatulate tip, the upper part decorated with a series of wispy lines. 14.2 cm long. Crozo de Gentillo (Lot), France. Magdalenian. The object may be functional or a representation of a plant or both. (70) is a blade resembling a modern letter opener. 13.3 cm long. Jouclas (Lot), France. Solutrean. Incisions along both sides at the base, perhaps for attaching a handle. Microscopic examination indicates that the sharp edges seem to have been polished by use. Both loaned by Logan Museum, Beloit College. Photos by R. White. (52,60)

Control of Heat and Light

Plant materials, especially wood, would have been important as fuel for cooking, heating, and light. Again, the excellent preservation at Lascaux indicates that certain species of trees and shrubs were sought especially as fuel for bonfires and torches and as wicks for the stone lamps used to provide light in the cave. The charcoal recovered from Lascaux comes primarily from juniper bushes (Leroi-Gourhan et al., 1979). Their wood provide long-burning aromatic fuel because of its high resin content.

Lamps were most often small slabs of limestone hollowed in the center to hold the fuel. Hundreds of these have been discovered from sites all over Europe. However, some of these lamps were very carefully prepared. Special kinds of stone obtained from distant regions were preferred. These special lamps (Figs. 71–75), which seem especially prevalent in painted caves, were often decorated with abstract or animal forms.

A recent analysis of several hundred stone lamps by Sophie de Beaune-Roméra (1983, 1985) of the University of Paris I indicates that Ice Age people employed some kind of animal fat in conjunction with a wick of moss or some other porous material. De Beaune-Roméra was able to reconstruct Ice Age lamps and to measure the amount of light given off. She concluded that the amount of light produced in a dark place would have created the light conditions that we associate with dusk. It should be pointed out that many stone objects that have been called lamps show no evidence of burning. Some of them may be grinding stones or mortars.

People relied primarily on cobble-lined fireplaces for warmth in cold weather (Fig. 83). These take a surprising number of different forms that vary from period to period and from region to region. They vary from very small (a few centimeters in diameter) to immense (as much as two meters in diameter). Some were dug into the ground; others were not. Some were lined with hundreds of cobbles, while others are rings of just a few cobbles. Where available, wood served as the primary fuel. However, bone scraps (which make excellent fuel because of their fat content) were also used, especially in periods and regions lacking abundant wood.

Fig. 71a,b Lamp of
ground red sandstone,
broken during excava-
tion and later pieced to-
gether. 17.2 cm long x
11.9 cm wide x 4.2 cm
thick. La Mouthe Cave
(Dordogne), France, Late
Magdelenian (?). (a): Up-
per surface of lamp. The
circular depression,
about 10 cm in diameter,
contained the fuel, prob-
ably animal fat with a
moss or bark wick. A few
engraved lines decorate
its margin. The tapered
end at right probably
served as the handle. Surface is coated with car-
bon. (b): Undersurface of
lamp, engraved with ibex
head in profile. Two
long, sweeping lines rep-
resent the horns. This
decorated lamp clearly il-
lustrates how ideas and
symbols were embedded
in even the most func-
tional of objects.
Musée des Antiquités
Nationales. Photos by
MAN. (75)

Fig. 72 Lamp of red sandstone. 22.3 cm long x 10.7 cm wide x 3.3 cm deep. Lascaux Cave (Dordogne), France. Early Magdalenian. A long incision and several shorter strokes decorate the top of the long handle. Many similar abstract designs are painted on the walls of Lascaux. This lamp is exceptional: several dozen others found in Lascaux were formed by pecking a crude depression in a rough piece of limestone. The sandstone used here occurs many kilometers to the northeast and would have taken some effort to acquire, unless collected during seasonal movement. Musée des Antiquités Nationales. Photo by MAN. (111)

Cooking included techniques for boiling. Upper Paleolithic sites often have pits full of cobbles fractured by temperature change in the same way that a glass bowl might break if removed from the oven and placed in the refrigerator. To boil water, people simply took red-hot cobbles from a fire and dropped them into a skin-lined pit full of water (Fig. 88).

Upper Paleolithic people developed strategies to reduce their fuel requirements as many people have done since (including ourselves in recent years). For example, in southwestern France as well as in the Ukraine, 75 percent of known Upper Paleolithic sites are located on south-facing slopes and at the bases of south-facing cliffs (White, 1985; Soffer, 1985). This familiar strategy is based on the recognition that, at European latitudes, south-facing embankments and cliff faces take on solar heat during the day and give it off slowly at night. Of course, south-facing places are also drier because of higher evaporation rates.

Settlement Strategies

Where people chose to live is an important reflection of strategies for survival and often tells us something about how they obtained necessary resources. Often, sites were so well situated that they were occupied, at least on a seasonal basis, for thousands of years. The

Fig. 73 Pecked and ground lamp of limestone, with linear incisions on the handle. 12.3 cm long x 8.3 cm wide. Le Grand Moulin (Gironde), France. Magdalenian. Musée d'Aquitaine. Photo by J. M. Arnaud, MA. (128)

Figs. 74, 75 Palette of red sandstone for grinding red ocher and three ocher crayons, one striated through use. 10 cm long x 7 cm wide x 2.6 cm thick. Crozo de Gentillo (Lot), France. Magdalenian. Ocher crayons 1.9 to 3.3 cm long. La Madeleine (Dordogne), France. Magdalenian. All loaned by Logan Museum, Beloit College. Photo by R. White. (66, 50)

Fig. 76 Mousterian and Upper Paleolithic site of La Ferrassie (Dordogne), France. It must have been a preferred location for prehistoric people over a long period: the thick accumulation of deposits, representing reoccupation of the site hundreds of times, is clearly visible. Earliest deposits date to about 50,000 years ago, the latest to 15–20,000. Remains of about a dozen Neanderthals, some purposely buried, were found at roughly the level of the ground surface in the foreground. Photo by R. White.

thickness of artifact-rich deposits at these sites can exceed 5 meters (1 feet) as shown (Fig. 76).

Proximity to reliable sources of water was one of the most important criteria in choosing a place to live (Fig. 78). More than 90 percent of all camps were located near springs or on the banks of rivers and streams.

Sites at the base of cliff faces sometimes turn out to be places where animals were driven over a cliff, to be butchered where they fell. In France, Upper Paleolithic sites often occur at shallows or fords in major rivers that were probably crossings for migratory animals such as reindeer (White, 1985). At one such site, the huge rock shelter of Laugerie-Haute on the Vézère River (Fig. 77). nearly complete reindeer skeletons were found between the rock shelter and the ford in the river. On a small rise nearby were found 21 pits dug into the ground, about a yard deep and the same across (Roussot, 1973). Stone tools in the bottom of these pits date them to the Solutrean. Several hypotheses await testing: animals may have been driven toward the pits, dug to serve as traps, into which they fell; the pits may have been foxholes, used by hunters to hide from approaching animals; or the pits may have been used for storing meat and other foodstuffs.

Fig. 77 The immense rock shelter of Laugerie-Haute follows the base of a tree-covered cliff (right half of photo). Covering more than 5,000 square meters (about 52,500 square feet), it lies adjacent to the shallows at middle left. Upper Paleolithic deposits are found right out to the Vézère River in the foreground. Photo by R. White.

Many sites seem to have been chosen because they provided a good view of the surrounding area (Fig. 79). Such vantage points would have been important for observing game animals and perhaps other humans. One thing is clear. People could live where they chose, on the basis of their decisions about the availability of necessary resources. Frequently, they chose caves and rock shelters (Fig. 81)—but only if the shelters were suitably located. For example, plenty of caves and rock shelters were never occupied because they faced north and were open to wind and rain. Clearly, people did not need to live in natural shelters such as caves and rock overhangs. Many of the most recently discovered sites are in open-air situations (Figs. 82,84), often miles from any cave or rock shelter. It seems that archeologists have perpetuated the "cave dweller" myth by looking for caves and rock shelters in their search for new sites. This approach has been the

Fig. 78 Magdalenian site of La Truffière (Dordogne), France, situated at the base of the cliff seen at right, includes a natural spring. Proximity to water was important in locating Upper Paleolithic camps. Photo by R. White.

easiest way of making new discoveries, as buried open-air sites are hard to locate. They must be sought by excavating in places that often show no signs of artifacts on the surface (Gaussen, 1979).

Architecture

It is in open-air sites that we find rich evidence for a variety of architectural features indicating clearly that Upper Paleolithic people were more than capable of sheltering themselves. Perhaps the most frequently found structures are pavements made of river cobbles carried from some distance away. These structures, often square or rectangular, were probably the foundations for shelters made from skins and wood or animal bones (Fig. 86). The foundation cobbles are almost always burnt, but Jean Gaussen (1979), an important pioneer in open-air research, has shown that they were burned before they were laid down, not after. (Half are burnt on their upper surface, half on their buried surface.) It seems that the builders heated the cobbles in a fire and then arranged them on the ground. Gaussen has suggested that they may have laid heated cobbles on frozen winter ground to form a secure platform. This precaution would have prevented a messy living surface from developing inside a shelter when human activities resulted in melting of the frozen ground underneath.

Fig. 79 View from the Late Magdalenian site of Font-Brunel (Dordogne), France. The view extends more than 10 kilometers (6.25 miles) and would have allowed monitoring movements of animals and other humans. Many Upper Paleolithic sites commanded such views of the countryside. Photo by R. White.

Fig. 80 Upper Paleolithic sites vary in their size and complexity. One of the more complex is the site of Reignac in the Vézère Valley. Visible here is a large cave about three quarters of the way up the cliff. Below that is a large rock shelter, visible middle left, into which is built the "modern" (16th century) fortress in the center of the picture. In the cave and rock shelter and on the slope below, there is rich evidence of human occupation from the Late Magdalenian up to the historic period. Structure at lower right is area of recent excavation by Roussot. Photo by R. White.

Fig. 81   In contrast to
large sites such as
Reignac, most Upper Pa-
leolithic sites cover very
little area. The small rock
shelter of La Metairie
(Dordogne), France, was
occupied during the
Aurignacian. Obviously,
rock shelters serve their
original purpose, even today. Photo by R. White.

Fig. 82 a,b   In recent years, it has become apparent that many Upper Paleolithic camps were situated far from any cave or rock shelter. The location of one such site on the interfluve, high above the Vézère River is shown in (a). A test excavation by the author was located between the two trees at center of picture. The result was the discovery of the Early Magdalenian site of Mala, part of which is shown in (b) with tools in place.

Fig. 83   A number of different kinds of structures inform archeologists about Upper Paleolithic people. The sophisticated use of fire is evident in this fireplace, which is being excavated in an Early Aurignacian level at Cueva Morín (Santander), Spain. Excavations by L. Freeman and J. Echegaray. Photo by L. Freeman, © Institute for Prehistoric Investigations.

Fig. 84   A living surface at the Late Magdalenian site of Pincevent (Seine-et-Marne), France. The debris of everyday life about 12,000 years ago is scattered on the floor. The site is a model of modern scientific excavation. Living surfaces have been carefully uncovered; even eggshells were found intact. Recording the location of even the tiniest object has permitted the drawing up of detailed site plans. Spatial arrangement of objects is as important to understanding prehistoric culture as the objects themselves. Photo by J. Vertut. Permission of Arl. Leroi-Gourhan.

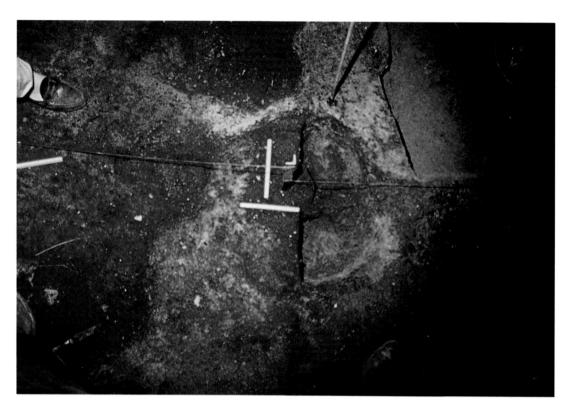

Even in caves and rock shelters, Upper Paleolithic people built additional shelters. For example, in the Spanish cave of Cueva Morín (Freeman, 1980), a "dugout hut foundation" and an alignment of postholes were uncovered in an early Aurignacian level.

Sometimes open-air Upper Paleolithic sites show spectacular and unexpectedly complex architectural features. They are especially noteworthy at recently discovered sites in the Ukraine (Gladkih et al., 1984), where the primary building material was the bones of woolly mammoths—and not just of one or two of them; in one case, bones of 95 of these 10,000-pound animals were used to make just one dwelling. At Mezirich, the site of this structure, there were three more such dwellings, composing a 15,000-year-old village (Fig. 85). Each dwelling had a different arrangement of the giant bones, which came from the skeletons of long-dead animals retrieved from the surrounding area by occupants of the site, not from animals they had recently hunted. The different arrangements of the bones forming each dwelling yielded very different architectural styles and may have

71

Fig. 85 (preceding page)   Remains of Upper Paleolithic mammoth-bone dwelling Number 4, about 15,000 years old, at Mezirich, Ukraine, on the Dnepr river. Over the foundation wall composed of various bones, the upper wall was formed by layering many bones of one type. This portion, facing southwest, was constructed of mandibles (lower jaw-bones) arranged in a her-ringbone design. Fifteen thousand kilograms (16½ tons) of bone were used for this dwelling. Four other such structures at Mezirich have different patterns of arrangement. Discovered 1976 by I.G. Pidoplichko. Photo by O. Soffer.

Fig. 86   Dwelling structure Number 2 from the Magdalenian site of Le Breuil (Dordogne), France. About 2 m. More than 100 cobbles form a pavement upon which seven tools (yellow tags) were found. Photo by J. Gaussen. (29)

Fig. 87   Accumulation of red deer bones, representing more than 50 individual animals. El Juyo (Santander), Spain. Magdalenian. Photo by L. Freeman. © Institute for Prehistoric Investigations.

Fig. 88 Two Gravettian living surfaces at Le Flageolet I (Dordogne), France. The large blocks are from the shelter's collapsed ceiling. A cobble fireplace (middle right edge) and a stone-filled boiling pit (lower edge of shadow at bottom right) are visible. Rock shelters are among the most geologically complex of all archeological sites, sometimes requiring up to 20 years to excavate a site that was occupied for only a few weeks. Excavated by Jean-Philippe Rigaud. Photo by R. White. Permission of J.-Ph. Rigaud.

reflected different esthetic requirements. University of Illinois archeologist Olga Soffer, who has worked at Mezirich, estimates that one of these dwellings alone would have taken 10 people 5 days to build. It is hard to know how many people lived at Mezirich but, if one calculates one nuclear family per dwelling, the site would have housed about 25 people.

Scheduling and Mobility

A fact of life for people who live by gathering plant and animal foods and hunting animals is the changing availability of these resources from season to season. Plant foods are generally available only in spring, summer, and fall. Some animals migrate and therefore come and go seasonally. Reindeer, for example, migrate twice a year in large herds.

The quality of an animal's meat or skin varies, even if the species is present year-round. During the fall reindeer migration, these animals are particularly attractive to people who hunt them because the animals are fat after a summer of feeding. Their coats are thick and lustrous in preparation for the winter ahead. During the spring migration, however, they are thin after having endured winter food shortage. Their coats are scraggly and often full of parasites such as warble flies.

Upper Paleolithic people had to schedule their activities accordingly. The best indication of this is that they occupied sites for only a portion of the year before moving on to other places and other resources. Their existence was a mobile one but by no means one of aimless wandering. They regularly returned to the same locations each year to hunt the same animals and gather the same plants. Frequently, they left behind small hoards, or caches, of materials such as high-quality stone or even spear-points, undoubtedly in anticipation of their return. For example, at Volgu in east central France, a cache of 14 large and finely worked Solutrean leaf-shaped points was found by workmen digging a canal. Likewise, at La Crouzette in the Vézère Valley (Blanc, 1953), excavations yielded many leaf-shaped Solutrean points, including one that was 36 centimeters (14 inches) long and weighed 1.52 kilograms (3.35 pounds).

Certain resources allow long-term human encampments if people know how to preserve and store food. There are recorded instances of Siberian reindeer hunters in May still living off mass kills made in late August (Popov, 1966). In many parts of Europe during the Upper Paleolithic, meat could easily have been stored over the winter simply by freezing it. Recent studies of animals killed in Upper Paleolithic sites show that, in some instances, people managed to live in the same location from September to March (Delpech, 1983). They occupied other sites for shorter periods, probably as short as overnight.

Nutrition and Health

How successfully did Upper Paleolithic people meet their nutritional needs? Recent research by Mary Ursula Brennan of New York University, using x-ray and other methods, indicates that during certain periods people were not faring very well. For example, about 40 percent of the skeletons recovered from Aurignacian sites in southern France show disruption in tooth enamel growth due to starvation or illness at some point in life. During the following Gravettian period, there is little such evidence. This variability leads us to conclude that continually changing environmental circumstances raised serious problems for human adaptation.

One of the most important factors in determining whether people will have enough to eat is the total number of people who must be fed in relation to available food resources. If the number of sites for each period is a good measure of relative population density, then there is clear evidence that when population density was highest, human nutrition was at its worst. The two periods with the most sites, the Aurignacian and the Magdalenian, have the highest incidence of growth disruptions.

An aspect of health among early *Homo sapiens sapiens* that is of far-reaching significance concerns longevity. The best evidence available (Heim, 1976; Thompson and Trinkaus, 1981) reveals that no

Figs. 89, 90   Two ivory needles with eye. Jouclas (Lot), France. Probably Soulutrean. First appearing in the Solutrean and abundant in the Magdalenian, such needles were produced by removing a small splinter from a larger piece of bone or ivory and then carefully smoothing it. The minute eye hole was drilled by a very fine stone perforator. Photo by R. White. (55, 56)

Neanderthals survived past their mid-forties. In contrast, at least some Upper Paleolithic people were living into their sixties (Skinner, 1981; Smith, 1977). Of course, this difference means that the social environment of modern humans would have included at least one more generation and that the knowledge transmitted to youngsters and available to the social group as a whole was much greater.

Clothing

The clothing that late Ice Age people wore is not preserved. However, many indirect forms of evidence allow us to draw inferences about it. One of the important innovations of the Upper Paleolithic was the eyed sewing needle (Stordeur-Yedid, 1979), which first appeared about 20,000 years ago, during the Solutrean. The needles are most frequently made of ivory or bone, and many are as small as large modern examples (Fig. 89,90). They have a tiny eye made by a very fine-pointed flint drill, or perforator. The "thread" used was probably animal sinew. Presence of such needles suggests that Upper Paleolithic people had well-tailored clothing, probably made from animal skins and somewhat like that of modern Eskimos and some American Indians. Of course, the fact that many stone tools show evidence of having been used to work hide further supports the idea that clothing was made from skins.

Occasionally, we find works of art that indicate clothing. One of these, found in the cave of Gabillou in the Isle Valley of France (Gaussen, 1964), seems to show a person, probably a woman, wearing what is interpreted as an anorak, or parka (Fig. 91). Also at Gabillou, a human, possibly a sorcerer, seemingly wears the skin of a bison with head still attached (Fig. 92). Perhaps the image indicates a form of ceremonial dress. Several of the 110 or so engravings of humans from the 15,000-year-old Magdalenian site of La Marche in France (Pales and Tassin de Saint-Pereuse, 1976) seem fully dressed in tailored clothing with cuffs and collars. How true to life these representations are is questionable, however, as primary sexual characteristics are represented despite what appears to be clothing. Sometimes the skeletons in burials are covered in beads that must

Fig. 91   Figure known as
"La femme à anorak"
(woman with parka) en-
graved on wall. Gabillou
Cave (Dordogne), France.
Magdalenian. Sex of the
figure is unclear and see-
ing the anorak requires
imagination. Photo by J.
Vertut.

have decorated the clothing of the dead individuals. One of the most remarkable pieces of such evidence concerning clothing comes from the 24,000-year-old Gravettian site of Sungir, one of the most northerly sites on the Russian plain, where three burials were found. O.N. Bader's careful excavation and description of these burials (1970) allows a clearer understanding of Upper Paleolithic clothing in eastern Europe. His description of one of the burials (Bader, 1964), that of a 55-to-60-year-old man, is worth reproducing here.

The bone ornaments consisted of polar fox fangs with apertures bored in them, two dozen bracelets in the shape of thin plates carved from the tusks of a mammoth, and beads made of the same material and sewn on the clothes in rows. Their distribution in the grave makes it possible to reconstruct, in general outline, the clothes which, in the case of these reindeer and polar fox hunters, were made of fur, of course. The people wore trousers and a shirt which had no cut in front and, hence, was put on over the head. There were, besides, some upper garments thrown over the corpse. There is good reason to assume that the trousers were stitched together with the leather footwear making up a single piece, as was the case with some later tribes of North Siberia and North America.

Clearly, the people of Upper Paleolithic Europe had at their command a body of technological knowledge as sophisticated as that of any modern hunting and gathering people. This knowledge was passed down through hundreds of generations of anatomically modern humans, who added to it and improved upon it. Such continuity and intergenerational communication cannot exist, however, without means of organizing people and production so as to ensure both social and biological reproduction.

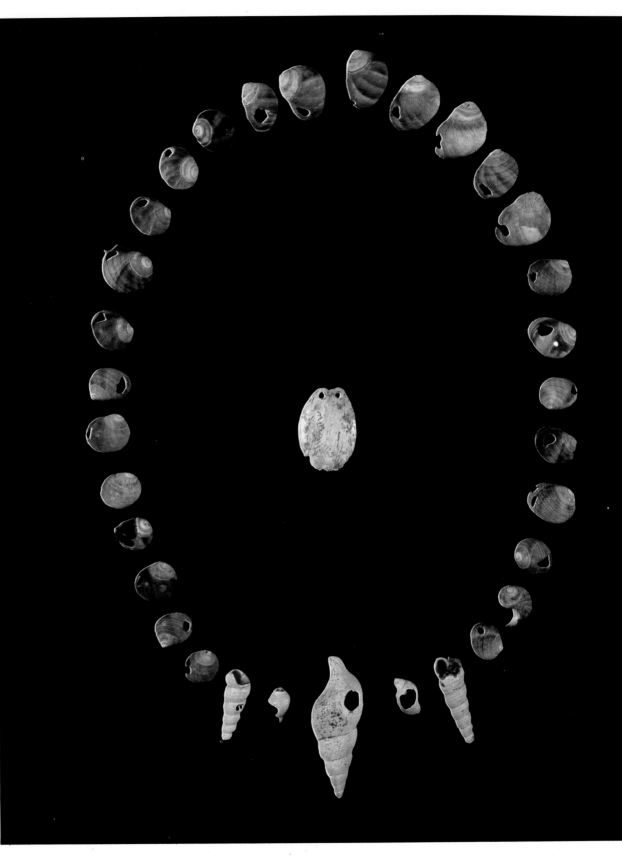

CHAPTER 3

The Social World

Fig. 93 Series of mollusk shells, perforated for suspension—reconstructed as a necklace—and a two-holed ivory pendant, associated with some of the first Upper Paleolithic burials to be discovered. Ivory pendant: 3.2 cm long. Cro-Magnon (Dordogne), France. Aurignacian. Musée de l'Homme. Photo by J. Oster, MH. (150, 146)

Introduction

While we have quite a firm knowledge of many aspects of Upper Paleolithic technology and economy, there exist many more gaps in our knowledge of how Upper Paleolithic societies were organized. Social anthropologists studying modern people can make direct observations about such social institutions as marriage and kinship relations. Because archeologists are restricted to the study of material objects left behind and the ways such objects are patterned in space, they have greater difficulty in drawing inferences concerning social behavior. Our knowledge of Upper Paleolithic societies is, necessarily, of a more general nature.

Group Size

The first basic question about Upper Paleolithic society is the size of Upper Paleolithic social units. One of the main ways that archeologists approach this problem is by determining the surface areas of the sites left behind by the people who occupied them. The equivalent in the modern world would be noting that cities have larger surface areas than towns and villages. It has been clearly shown among modern hunting and gathering peoples that surface area of a settlement roughly reflects the number of people who live there (Yellen, 1977) and how long they stay.

Figs. 94–103 The Magdalenian site of Laugerie-Basse (Dordogne), France, was apparently a center of artistic activity and has yielded several hundred examples of portable art.

Fig. 94 Sculpted horsehead profile adapted to the natural contours of a tine of reindeer antler. 7.4 cm long. Laugerie-Basse (Dordogne), France. Middle Magdalenian. Musée de l'Homme. Photo by J. Oster, MH. (168)

Fig. 95 Two cervids sculpted in relief on broken semicylindrical antler rod; the front animal may be a young stag growing its first set of antlers. 7.4 cm long. Laugerie-Basse (Dordogne), France. Middle to Late Magdalenian. Musée de l'Homme. Photo by J. Oster, MH. (164)

Fig. 96 Two horse-head profiles engraved on piece of rib broken at both ends. 9.5 cm long. Laugerie-Basse (Dordogne), France. Middle to Late Magdalenian. The nose of the first horse has been broken away and its facial features are less bold than those of the second; its mane is remarkably detailed by use of fine incisions. Musée de l'Homme. Photo by J. Oster, MH. (163)

When we estimate surface area of Upper Paleolithic sites, we find that there is a wide variety of site sizes for any given period. This finding seems to indicate that there were different kinds and sizes of social units. Among modern groups such as Eskimos, group size fluctuates seasonally, with at least one period during the year when large numbers of people aggregate (Mauss, 1904–1905), usually around some abundant resource. Such aggregation cannot be sustained, however: first, because the environment cannot support such dense concentrations of consumers for long and, secondly, because hunting and gathering societies seldom have the social institutions to resolve the conflicts that occur when many humans come together (Lee, 1979). The primary means of resolving conflicts is what anthropologist Richard Lee calls "voting with your feet"— disbanding and moving away from the social intensity of the aggregation. It can therefore be hypothesized that the many small and few large sites evident for the European Upper Paleolithic reflect a social pattern of fluctuation between small and large social units (although there was probably regional variation linked to different resource bases).

Indirect evidence exists to support this hypothesis. First, many of the largest sites appear to be located at places suitable for efficient exploitation of seasonally abundant resources, such as migrating reindeer or fish. In other words, there would have been a temporary resource base large enough to support many people. The site of Laugerie-Haute, mentioned earlier, is one such example.

Secondly, the largest sites (even when only a small area has been excavated) commonly yield many more art objects than smaller sites. For example, the very large Magdalenian site of Laugerie-Basse, next door to Laugerie-Haute in the Vézère Valley of southwestern France, has produced several hundred small portable art objects (Figs. 94–103), while many small sites produce few or none. Among modern hunting and gathering peoples, it is clear that much of the artistic endeavor takes place within the socially charged context of seasonal aggregation, when most ceremonies such as marriage and initiation to adulthood occur. It is also frequently a time of storytelling and

Fig. 97 Deer head sculpted on a piece of reindeer antler, possibly the middle section of a spear thrower. 7.4 cm long. Laugerie-Basse (Dordogne), France. Middle to Late Magdalenian. Musée de l'Homme. Photo by J. Oster, MH. (179)

intense ritual in which "art" objects play a key role. It appears that this pattern may have been true of certain periods of the Upper Paleolithic as well.

Thirdly, the analysis of human activities in the rock shelter of Flageolet I (Fig. 88) in the Dordogne Valley indicates the presence of only one socioeconomic unit, rather than a clustering of several self-sufficient units such as we usually find when large numbers of hunter-gatherers come together. Françoise Delpech and Jean-Philippe Rigaud (1974) have shown that the Gravettian occupants divided the rock shelter into complementary work areas for butchering, cooking, bone boiling, marrow extraction, and dumping of garbage. This seem to be the work of a single household that organized itself according to the available space of the shelter (Simek, 1984).

Fig. 98 Engraved rib fragment showing rear of one animal, a complete bovid, and the head of another animal. 20 cm long. Laugerie-Basse (Dordogne), France. Middle to Late Magdalenian. The engraved lines extending onto the broken surfaces at either end indicate that the engravings were done when the bone was already in its present form. Musée de l'Homme. Photo by J. Oster, MH. (178)

Fig. 99   Head of a cervid
engraved on a bone frag-
ment and facing away
from a flowerlike image.
Laugerie-Basse (Dor-
dogne). France. Middle
to Late Magdalenian.
Musée de l'Homme.
Photo by J. Oster, MH.

Fig. 100   Flat section of
broken reindeer antler
engraved with bison
head in profile. 7 cm
long. Laugerie-Basse
(Dordogne), France. Mid-
dle to Late Magdalenian.
Musée de l'Homme.
Photo by J. Oster, MH.
(177)

Long-Distance Social Contacts

Another important social question is the area over which social ties were developed and extended. The answer is surprising to those who conceive of Upper Paleolithic societies as small, self-contained units. Certain Upper Paleolithic sites in the Ukraine contain seashells found only in the Mediterranean. Baltic amber has been found in sites in southern Europe. Sites in inland France and Spain contain shells from the Mediterranean (Figs. 104,105), the Atlantic, the English Channel, and fossil shell beds hundreds of miles distant. Late Upper Paleolithic sites in Poland contain obsidian and jasper from several hundred miles away (Schild, 1976). While finds so far from the original sources could be explained by actual movements of people (Bahn, 1977, 1982), it is more reasonable to suppose that these goods moved in hand-to-hand exchange as has been observed in many modern hunting and gathering peoples (Weissner, 1982; Sharp, 1952). Shells often have special social value and even magical properties for people removed from the sea and may have been eagerly sought.

Fig. 106 Partly flexed
burial before removal
from the Magdalenian
site of Cap-Blanc (Dor-
dogne), France. Part of a
stone structure built
around the dead person
is still visible. Photo courtesy of Field Museum of
Natural History, Chicago.

In the modern world, we tend to think of exchange, or trade, as a purely economic transaction. But in most small-scale societies, it operates as a vehicle of social obligation. As the anthropologist Marcel Mauss (1923–1924) pointed out, the acts of giving and receiving set up a never-ending series of obligations—much like the exchange of gifts at Christmas. Obligations are social bonds capable of tying together different social groups. It is perhaps in this context that shells and flint moved across the ancient European landscape.

Social Divisions

No known human societies are 100 percent egalitarian. Even the least hierarchical societies known to anthropologists are subdivided along lines of age, gender, and personal achievement. In archeology, information about this internal subdivision is most frequently recovered from burials. Within limits, all modern human societies express a person's social position in ceremonies, offerings, or structures at the time of his/her death. Observed differences in the way the dead are treated within the same society can reasonably be assumed to reflect the different statuses or roles that they filled in life.

The European Upper Paleolithic shows a wide range of burial treatment (Fig. 105) within given regions (Binford, 1968; Harrold, 1980; Pike Tay, 1984). There is a staggering diversity in such features as the position of the body (flexed, extended, highly flexed), ritual treatment of the corpse, nature and quantity of grave goods accompanying the corpse, structures associated with the burial, number of corpses interred in the same grave, and the way in which the body was clothed or decorated.

Differences often exist between regions, perhaps suggesting different social traditions in mortuary treatment. However, the opposite can also be true. A 55-to-60-year-old man at the site of Sungir in the USSR (Bader, 1964; Soffer, 1985) was given burial treatment similar to that of a young girl at La Madeleine (Capitan and Peyrony, 1928) in southwestern France: wrists, elbows, knees and ankles were decorated with bracelets or multiple strings of beads, and the skull also had a band around it. The only difference is that at Sungir the

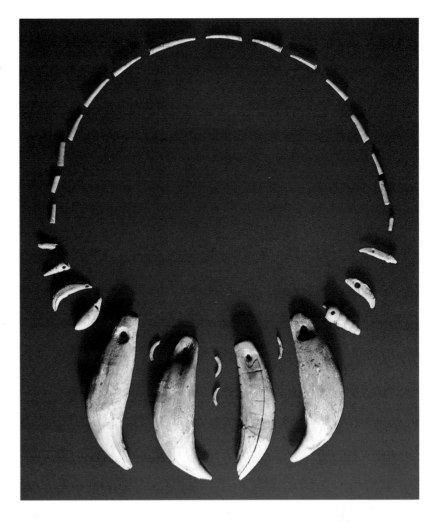

Fig. 107 Necklace composed of pierced teeth and various shells. Dentalium shells from the Atlantic coast 150 kilometers (95 miles) distant form the upper half. Rocher de la Peine (Dordogne), France. Late Magdalenian. The four large teeth are of bear and lion (second from right). The lion tooth has been incised on all sides. Logan Museum, Beloit College. Photo by R. White. (67)

Fig. 108 Series of 137 stone, bone, and ivory beads and pendants—probably clothing beads and pendants—reconstructed in 1925 by A. Pond as a necklace. Abri Blanchard (Dordogne), France. Aurignacian. Logan Museum, Beloit College. Photo by R. White. (63)

bracelets were made of mammoth ivory beads, while at La Madeleine they were made of Dentalium shell segments.

Many of the adornments worn by the dead may simply reflect everyday decoration in life (Fig. 107). Present-day peoples wear various ornaments, at least some of which convey wealth or social standing. The 30,000-year-old skeletons discovered at the site of Cro-Magnon in the late 1860s apparently had necklaces of shells (Fig. 93)— perhaps accompanied by ivory pendants. A necklace (Fig. 108) discovered at Rocher de la Peine on the Vézère River was associated with the skeletal remains of a youth and was made of a variety of pierced teeth (notably bear and lion canines) and shells. It is generally assumed that such bodily adornment communicated one's social standing (Figs. 109–119).

Fig. 109   Carved bone
pendant with lower ap-
pendage bordered by in-
cisions, one of the earliest
body ornaments known.
3.7 cm long. Arcy-sur-
Cure (Yonne), France.
Aurignacian. Collection
A. Leroi-Gourhan. Photo
by J. Vertut. Permission
of Arl. Leroi-Gourhan.
(36)

Fig. 110   Carved mam-
moth-ivory "pendant,"
representing a human
figure, probably a fe-
male, with head and
hairline clearly visible.
5.1 cm long. Abri Cellier
(Dordogne), France.
Aurignacian. The signifi-
cance of paired incisions,
which are frequent on
early Aurignacian art ob-
jects, remains unknown.
Logan Museum, Beloit
College. Photo by R.
White. (45)

Fig. 111 Bone pendant decorated with incised bounded concentric circle design. 5.8 cm long. Saint-Marcel (Indre), France. Middle Magdalenian. Musée des Antiquités Nationales. Photo by MAN. (74)

Fig. 112 Sitting bear sculpted of stone with a central perforation and clearly visible facial details—probably a large pendant. 5.8 cm high. Isturitz (Basses-Pyrénées), France. Middle Magdalenian. Musée des Antiquités Nationales. Photo by MAN. (116)

Fig. 113 Pendant of mammoth ivory, representing a fish or seal, from the "necklace" in Fig. 108. Abri Blanchard (Dordogne), France. Aurignacian. Punctuations cover the back and continue onto the sides. The muzzle, originally pierced from top to bottom, has been broken. About 4 cm long. This and a phallus sculpted of ivory from the same site are the oldest pieces of portable figurative art in France. Logan Museum, Beloit College. Photo by R. White. (63)

Especially in the later phases of the Upper Paleolithic, it is not unusual to find the dead treated in what is by our standards a ghoulish fashion. At the Magdalenian site of Ofnet in Germany, more than 30 individuals, most of them reportedly adult females, had been decapitated (either while living or after death), and their skulls had been carefully placed in two broad, shallow pits inside a rock shelter. Their cervical vertebrae showed marks made by stone blades. All of the skulls faced in the same direction. At the site of Le Placard in southwestern France, about a dozen skullcaps had been made into bowl-shaped objects. Other bones found in the site show clear evidence of stone-tool cut marks.

Initially, this kind of treatment seems to be clear evidence for brutality and aggression. However, it may just as well mean the opposite—respect. Many societies have prescribed ways of treating their dead (Ucko, 1969) that include burning them, removing all of their flesh, and even hanging them in trees. It is entirely possible that the treatment afforded any particular individual depended upon his or her social position in life.

Many stone and bone tools carry elaboration that cannot be construed in purely functional terms. Bone and antler spear-points have animal and abstract images engraved upon them. These have been shown to vary between sites and regions and may be emblematic of different social groups or even of different hunters. Certain stone tools or weapons at certain periods are so elaborate that they have been interpreted as status or craft symbols. This is especially true for many of the leaf-shaped points or knives of the Solutrean period.

Fig. 114   Feline sculpted of mammoth ivory. Coat is indicated by series of punctuations. Ear and mouth are clearly indicated. 9 cm long. Vogelherd, Germany. Aurignacian (32,000 years old). Original loaned by Institut für Urgeschichte, University of Tubingen. Photo of cast by D. Ponsard, Musée de l'Homme, MH. (30)

Fig. 115   Woolly mammoth sculpted of mammoth ivory. Eye and trunk clearly indicated, as is the profile. Note same use of x's as in 114. Original loaned by Institut für Urgeschichte, University of Tubingen. Photo of cast by D. Ponsard, Musée de l'Homme, MH. (31)

Often, images that appear to be humans dressed as animals are taken as literal portrayals of shamans or sorcerers. The previously mentioned bison/human figure from Gabillou is one of the clearer examples (Fig. 92). This inference is debatable; however, if it is correct, it identifies a special social role for certain individuals, based on their spiritual or magical powers. Likewise, one might argue that if these and other images were created by a few expert craftspeople, they may well have had an esteemed social role within Upper Paleolithic society, especially given the extremes to which they had to go to paint, sculpt, and engrave deep underground.

It is often suggested that certain of the decorated caves were places where youngsters were initiated into adulthood. In particular at Le Tuc d'Audoubert (Begouen and Breuil, 1958), a deep cave in the French Pyrenees, footprints supposed to be those of youngsters have been interpreted to suggest dancing in a small chamber off the main corridor, well over one-half mile from the cave entrance (Fig. 120). Two spectacular bison sculpted in clay and dating to about 15,000 years ago stand deeper in the cave and past this side chamber (Fig. 121). A third sculpted figure, often interpreted as a juvenile bison, was present but in a very poor state of preservation. Although the sexes of the two larger bison are not explicitly shown, this may well be a mating scene constructed for ritual purposes associated with initiation ceremonies.

If this scenario is reasonably correct, then age-based social status is implied. However, the problem with such scenarios is that they are based on observations made on modern hunting and gathering societies. They use the present to understand the past instead of using the past to understand how the present came to be. This is a serious problem with many of the reconstructions of Upper Paleolithic society.

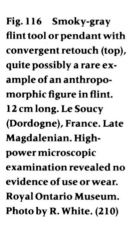

Fig. 116 Smoky-gray
flint tool or pendant with
convergent retouch (top),
quite possibly a rare ex-
ample of an anthropo-
morphic figure in flint.
12 cm long. Le Soucy
(Dordogne), France. Late
Magdalenian. High-
power microscopic
examination revealed no
evidence of use or wear.
Royal Ontario Museum.
Photo by R. White. (210)

Fig. 117 Reindeer antler
object covered with red
ocher and having linear/
geometric engraving, of-
ten compared to "bull-
roarers" of Australian ab-
origines, which make a
reverberating sound
when swung on a cord.
16 cm long. Roche de
Lalinde (Dordogne),
France. Musée des
Antiquités Nationales.
Photo by MAN. (96)

Fig. 118 Wedge-shaped object, probably a pendant, engraved with geometric designs that are often considered schematic representations of fish. 7.3 cm long. Laugerie-Basse (Dordogne), France. Middle to Late Magdalenian. Musée des Antiquités Nationales. Photo by MAN. (81)

Fig. 119 Broken bone pendant with engraving of bison and schematic humans, usually considered a hunting scene but probably having more symbolic significance. The image is so similar to another from the same period found at Chateau des Eyzies (Dordogne), France, that a widely shared story is implied. Raymonden-Chancelade (Dordogne), France. Magdalenian. Musée du Périgord. Photo by MP. (192)

Fig. 120 Human foot prints in clay. Niaux Cave (Ariège), France. Middle Magdalenian. Photo by J. Vertut.

Fig. 121   Two 15,000-year-old bison modeled in clay. Third figure, probably a calf, was at feet of animal at right. About a mile underground, at end of Le Tuc d'Audoubert Cave (Ariège), France. Middle Magdalenian. Front bison: 60 cm long. Photo by J. Vertut.

CHAPTER 4

The World of Ideas and Symbolic Expression

Fig. 122 Cave ceiling polychrome painting of bison in different postures, some standing and some possibly rolling on the ground. Altamira (Santander), Spain. Magdalenian. Photo by J. Vertut.

Introduction

It is safe to say that the most widely known but least understood aspect of Upper Paleolithic life is the extraordinary art produced during the period. Nearly 200 caves bearing wall paintings and engravings are now known from southwestern Europe, especially from France and Spain. In addition, perhaps 10,000 sculpted and engraved objects are known from across Europe and as far afield as Siberia and from several locations in Africa. Upper Paleolithic people mastered a wide range of artistic media: stone, bone, antler, ivory, wood, paint, clay, sound, and movement. The visual art ranges from naturalistic to abstract but encompasses relatively few themes. Animal images are clearly the most important of these; plants and landscape features appear less often. Contrary to a common misconception, humans or humanlike forms (including painted human handprints) are numerous, with approximately 1,000 examples known (Vialou, 1985). In addition, a wide variety of nonrepresentational signs exists. We are unable to interpret these signs because they undoubtedly comprise a body of arbitrary symbols just as written words do for us.

Evolutionary Implications of Art

One of the defining characteristics of modern members of *Homo sapiens sapiens* is that they experience their world through complex

103

Fig. 123 Fragment of reindeer antler engraved with barbed sign and punctuations assumed to be symbolic images for which the system of meaning is unknown. Laugerie-Basse (Dordogne), France. Middle to Late Magelalenian. Musée de l'Homme. Photo by J. Oster, MH.

conceptual frameworks that are often mutually unintelligible between cultures. In confronting cultures foreign to us, the most confusing material is often the body of sensory images that we in the modern Western world subsume under the term "art." While we tend to think of art as evocative, that which is evoked depends heavily upon participation in a mutually understood system of meaning—in other words, a shared body of ideas, conceptions, and experiences. Far from being a universal language, art is culture-bound.

It is commonplace for anthropologists to scratch their heads in confusion when first coming in contact with different systems of meaning (Figs. 123–125). Gradually, their head-scratching is reduced as they come to understand from their informants the rationale behind what they have observed. For archeologists encountering an Ice Age painted cave, there are no informants. They have all been dead for thousands of years. Therefore, our head-scratching is bound to go on for some time.

However, this very fact is of utmost significance. The fact that Ice Age humans, using visual media, were conveying ideas so complicated as to totally confound us indicates that the nature of the human adaptation had begun to change. In a very real way, humans produced their own world, a symbolic world that included imaginary creatures and deities. As we shall see, even the animal images conformed, in location and form, to a set of ideas and conceptions that are purely cultural.

Fig. 124 Dots in quadrangular arrangement marking a chamber entrance at Le Combel, Pech-Merle Cave (Lot), France. Such groupings are often found at the ends of painted galleries and where the cave turns suddenly. One of the many examples of "nonanimal" cave art images. Photo by J. Vertut.

If different *contemporary* cultures can have mutually unintelligible conceptual bases, it is certainly reasonable to suppose that major changes in systems of meaning and belief occurred over the 25,000 years of the Upper Paleolithic. It is often assumed that somehow Upper Paleolithic art can be viewed as a unified phenomenon and that the same interpretive frameworks can be applied to both the Aurignacian and the Magdalenian. This is an assumption that appears unwarranted.

The most recent attempt to provide a developmental chronology for the art is that of André Leroi-Gourhan (1967), the French anthropologist whose work in the 1960s quickly replaced the earlier framework of the Abbé Henri Breuil, a pioneer in examining, documenting, and analyzing Ice Age art. While Leroi-Gourhan seemingly demonstrated long-term continuity in art styles and certain

organizational attributes, he also pointed out major differences through time in location, media employed, subject matter, and form in Upper Paleolithic art. Part of the complexity of human symbolic systems lies in the fact that the same image can carry several meanings at the same moment and at different times. We must avoid the facile assumption that an engraved horse in the Gravettian is the symbolic equivalent of an engraved horse in the Magdalenian, or that a barbed sign in Cantabrian Spain meant the same thing as it did in the Dordogne or Italy. Let us fine-tune our observations by looking at what is known of chronological evolution and change in Upper Paleolithic art as well as the spatial organization of painted caves. Then we can proceed to examine the context (both physical and semantic) in which the art was being produced, used, and observed.

Fig. 125 a,b   Two sides of pierced bone baton with engraved decoration, nonanimal in nature. La Colombière (Ain), France. Late Magdalenian. Musée de Brou. Photo by MB. (144)

The Evolution of Upper Paleolithic Art

One of the more uncertain aspects of Upper Paleolithic art is its chronology. Dating individual works is difficult, especially in the case of wall paintings, which cannot be assumed to relate to the objects in the soil at their base. Theoretically, the situation is less tricky for portable objects such as small stone slabs and engraved bone and antler objects. However, the majority of art discoveries occurred before 1930, when radiocarbon dating did not exist and little precision was employed in recording stratigraphic levels. The lack of detailed information means that we have solid dates for perhaps only 10 percent of the known portable art objects.

This lack has resulted in several attempts to create relative chronologies for the art, based upon stylistic similarities and differences in the paintings, engravings, and sculptures themselves. The history of attempts at chronology will not be explored here, but the interested reader is referred to Margaret Conkey's discussion (1981). Leroi-Gourhan (1967) used the small sample of dated portable art objects as a basis for describing the evolution of art styles. He then dated images painted or engraved on cave walls according to their stylistic similarity to the dated objects. There is controversy over the accuracy of this approach and perhaps even more over his basic conclusion that Upper Paleolithic art underwent a continuous development over a 25,000-year period, a development independent of technological evolution in many ways. In other words, he concluded that there is an underlying artistic and symbolic continuity behind major changes in material culture.

Leroi-Gourhan clearly recognized that there were major quantitative discontinuities in artistic production, however. Certain periods and certain sites have yielded art objects by the hundreds; others show virtually no evidence of artistic activity. The art of some periods is deep underground; the art of others is in the open. This uneven pattern means that generalizations must be built from a set of complex, detailed, and often contradictory observations. Objects and representations must be carefully and accurately dated, and their locations within living sites and caves must be carefully recorded.

Unfortunately, at least 80 percent of the known art works were discovered prior to the general use of modern scientific techniques. Because the sample of well-dated art objects is so small, the possibility still exists that a single well-excavated, well-dated new discovery may change our notions of chronological development and variation.

Mousterian

The earliest objects that we are willing to call art are firmly dated to the late Mousterian. In some instances, a glimmering of artistic sensitivity is evidenced simply by an interest in curious forms; fossils, for example, are occasionally encountered in Mousterian sites. A smoothed and polished plaque purposefully shaped from mammoth ivory was found in the Mousterian site of Tata in Hungary. La Ferrassie in France has yielded small blocks of limestone with pecked out cup marks. But these examples indicate the limit of artistic expression in the Mousterian materials thus far recovered. There are no representations of animals. There are no engravings on bone— with the possible exception of examples from the Spanish site of Cueva Morín, recently excavated by Leslie Freeman and Jesus Echegaray (Freeman, 1983). Even these are but simple linear series of markings, however.

Châtelperronian

Artistic production became scarcely more complicated in the earliest phase of the Upper Paleolithic, known in western Europe as the Châtelperronian. This near absence of art has led some paleoanthropologists to suggest that the Châtelperronian is really just the last phase of the Mousterian, a position fortified by the discovery of the Neanderthal burials at Saint-Césaire. Recently, a sandstone plaquette bearing an uninterpretable engraving was found by Guy Mazière and Jean-Paul Raynal (1983) at the site of Grotte du Loup in the Corrèze region of southwestern France. In excavations at Arcy-sur-Cure in central France, André Leroi-Gourhan (1967) found a small number of objects bearing parallel incisions. He took these as perhaps the earliest evidence for rhythmic arrangements and an interval scale that was the prototype for the ruler, the calendar, and even the musical scale. This site also produced one of the most remarkable findings in the Châtelperronian—hundreds of pounds of natural clay pigments in a wide variety of colors, suggesting that perishable surfaces such as animal or human skin and wood may have been painted. However, no painted or engraved caves have been dated to the Châtelperronian.

Aurignacian

It is with the Aurignacians that we see the first unmistakable human and animal figures and the first evidence of music. Aurignacians produced a surprising diversity of graphic forms (Figs. 126–133). There are fragments of bird bone with carefully spaced incisions (Fig. 132,133). Some may have been used for notation or keeping track of events, as suggested by Alexander Marshack (1972). There are engravings and sculptures in the form of sex organs, most often female (Fig. 127,128). There are limestone blocks with simple, often fragmentary, animal forms (Delluc and Delluc, 1978). There are ivory plaques with series of punctuations. Surprisingly, however, some of the earliest surviving art objects are tiny three-dimensional animal sculptures (Figs. 114, 115) in mammoth ivory (Hahn, 1972). These are most numerous in Germany, but at least one example (Fig. 113) is known from France (White, 1986). The complexity of these sculptures leads us to ask whether there might have been prototypes in nonpreservable media such as wood and clay. In some instances, engraved blocks retain traces of paint.

Perhaps the most unexpected object that has survived from the Aurignacian is a wind instrument, frequently described as a flute, from the Abri Blanchard in southwestern France. It has six holes, two on the top and four on the underside. The complexity of its sound capability makes clear that the instrument was not just a whistle or animal call. The flute, which is about 32,000 years old, has been the subject of an unusual experiment by archeologist Mark Newcomer of the University of London. An expert in the reproduction of Upper Paleolithic artifacts, he fabricated an exact replica (Fig. 136) of the

Fig. 126 Engraved salmon on rock shelter ceiling. About 105 cm long. Abri de Poisson (Dordogne), France. Probably Aurignacian. Alterations surrounding fish are from attempts in 1914 to remove the image for sale to a museum. Photo by R. White.

Figs. 127, 128 Two large limestone blocks with engraved images interpreted as vulvae. Dozens of such images are known from Early Aurignacian sites in southwestern France and are among the oldest art objects on earth. Abri Cellier (Dordogne), France. Aurignacian. Musée des Antiquités Nationales. Photo by MAN. (91, 92)

Fig. 129 Long bone splinter engraved with paired incisions, a frequent element of Aurignacian art objects. 16.1 cm long. Abri Cellier (Dordogne), France. Aurignacian. Logan Museum, Beloit College. Photo by R. White. (39)

Fig. 130 Engraved bone plaque with punctuations and incisions similar to those in Fig. 129. 9.7 cm long. The punctuations may have been done by different tools, changed at intervals that correspond to lunar periods, but their arrangement may simply be decorative or symbolic or both. Abri Blanchard (Dordogne), France. Aurignacian. Musée des Antiquités Nationales. Photo by MAN. (85)

Fig. 131 a, b Two faces of a mammoth-ivory plaque with complex punctuations and incisions—a common form of decoration during the Aurignacian. 10.1 cm long. Abri Lartet (Dordogne), France. Aurignacian. Musée des Antiquités Nationales. Photo by MAN. (70)

Fig. 132, 133 Two fragments of bird bone incised with linear marks on three sides, function unknown but often called "hunting tallies." Smaller specimen appears to be complete, with smooth, polished ends. Left: 7.1 cm long; Right: 6.3 cm long. Abri Cellier (Dordogne), France. Aurignacian. Logan Museum, Beloit College. Photo by R. White. (38, 46)

Blanchard flute. The professional flutist Jelle Atema then played it and thus provided some rough approximation (biased by modern musical traditions) of the sounds that must have echoed around the rock shelter of Blanchard over 300 centuries ago. Another flute (Fig. 135), unfortunately broken, was discovered at the somewhat later (Gravettian) site of Pair Non Pair, also in southwestern France. The flutes are contemporary with the earliest figurative art; their presence indicates quite clearly that music was part of the artistic and symbolic environment that people had created for themselves. Even sounds were controlled and bestowed with meaning.

Gravettian

The Gravettian period, which followed the Aurignacian, brought an increase in complexity of art forms. At the site of Prêdmost in Czechoslovakia (Klima, 1966), numerous fired clay figurines, many in the form of humans, are perhaps the earliest evidence for ceramic art. Wall paintings exist for the Gravettian, and many more have probably not survived. Often in rock shelter sites, blocks that have apparently fallen from the ceiling bear traces of red and black coloring. An important theme at some Gravettian sites is the negative human handprint (Fig. 137), produced by brushing or blowing pigment around a hand held flat against a rock surface. The most impressive site featuring such prints is the cave of Gargas in the French Pyrenees, where A. Salhy (1963) has documented 217 handprints along with other possible traces. One of the most puzzling characteristics of the Gargas hands is that all except 10 have fingers missing. Early on, Cartailhac and Breuil hypothesized that the missing fingers resulted from ritual mutilation, a practice known from modern societies. Salhy, in a closely argued hypothesis, suggested that missing fingers resulted from disease and infection. Leroi-Gourhan (1967) suggested that fingers only appeared to be missing, having been held back against the palm of the hand. He went so far as to propose that a kind of gestural language or code was involved.

Leroi-Gourhan has remarked that deeply incised engravings (Figs. 138–140) are common and found in habitation sites rather than

deep underground toward the end of the Gravettian. For him, these
represent the prototype for the subsequent emergence of bas-relief
sculpture. An excellent example of such a figure was found on a large
block that had fallen from the ceiling in the Labattut rock shelter in
southwestern France (Fig. 138). The form—that of a highly stylized
horse— may have been engraved in broad and deep lines on the
block's surface after the block fell. The back of the horse conforms
well to the line of fracture, as if the fracture line had confined the field
available to the artist. It may have even suggested the contours of an
animal to the artist. This is a common feature of the art that Margaret
Conkey (1981) describes as "iconic congruence."

Few utilitarian tools were decorated in the Gravettian. The best
example is a spear-point with an engraved horse from the site of
Isturitz in the French Pyrenees (Fig. 134).

By far the most distinctive art objects in the Gravettian are
the female statuettes (Figs. 141–145), which have come to be known
as Venus figurines. These are found throughout Europe, often
showing conventional regional differences. They are sculpted from a
variety of materials, including ivory, steatite, and calcite. Certain
examples from eastern Europe were modeled in clay. The figurines
range in style: some, such as the magnificent Venus of Lespugue, are
highly exaggerated anatomically; others, such as the Venuses of

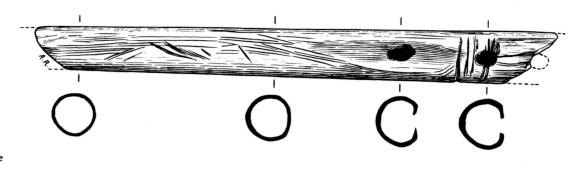

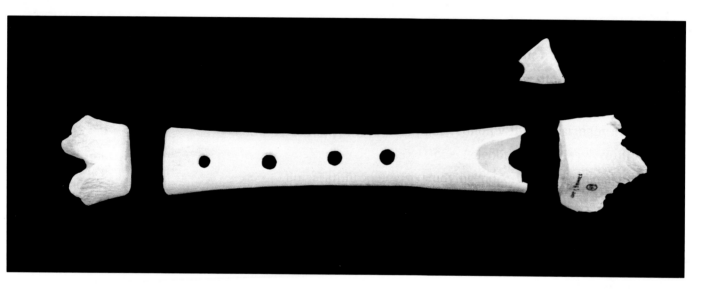

Brassempouy, are more moderately proportioned. Some appear pregnant; many do not. Seldom represented are facial features and lower legs (which often end in points). Henri Delporte (1979) has suggested that this style may have allowed the figurines to be stood upright in the ground. Breasts are most often large and pendulous, and buttocks are generally pronounced. In some cases, the pubic triangle is indicated. One remarkable specimen, the Venus of Monpazier, exhibits a very detailed vulva.

While the Venuses often take the form of small figurines, several examples are engraved or sculpted on large limestone blocks. Such is the case with the well-known "Femme à la corne" (Woman with Horn) from the site of Laussel in the Dordogne region of southwestern

Fig. 137 Negative image of human hand painted on cave wall probably by blowing paint through a tube. Pech-Merle (Lot), France. Magdalenian. Human handprints became important in art during the Gravettian, although this particular example is later. Photo by J. Vertut.

France (Fig. 148). Partially sculpted in bas-relief, the figure was also covered with red ocher. This relief figure and others associated with it deserve a closer look because they provide one of the few examples where a number of such objects were clustered in what may have been a purposeful arrangement. Unfortunately, the excavations took place long ago, and available information is not as precise as we might wish.

The site of Laussel occupies two contiguous rock shelters; one is about 450 feet long, and the other is about 65 feet long. They are separated by a gulley and overlook the River Beune, a tributary of the Vézère. The major excavations at Laussel were carried out by G. Lalanne between 1908 and 1914 (Fig. 146). They revealed a complex site with levels dating to the Mousterian, Châtelperronian, Aurignacian, Gravettian, and Solutrean. Solid evidence dates the female representations to the Gravettian.

The famous "Femme à la corne" was sculpted on a huge block (Fig. 147) that had fallen from the ceiling of the rock shelter and faced its intact wall. She is about 17 to 18 inches high and faces the viewer, although some interpret the figure as having her head turned to her right with her hair falling on the left shoulder. Her left hand, with four fingers visible, is placed on her abdomen; her raised right hand holds what is frequently interpreted as an incised bison horn. Red ocher was applied around the head and over most of the upper body.

A similar figure known as the "Venus à tête quadrillée" (Venus with Gridlike Head), was sculpted on a limestone block and was found nearby (Fig. 149). About 15 inches high, it also has the characteristic large abdomen and breasts and the raised hand (in this case, the left hand) holding something. The right hand hangs at her side and seems also to hold something. The head is covered in a grid similar to those seen on Venus figurines known from Italy and Austria.

A third female in the group, about 10 inches high, has similar proportions; unlike the other two, she has her right hand extended but not raised. She holds a curved object that is often interpreted as a skin or animal stomach for holding water. Unfortunately, this piece was sold to a Berlin museum by an unethical workman and was subsequently destroyed during World War II.

Fig. 138   Engraved horse
on massive limestone
block, in a style charac-
teristic of the Gravettian.
The deep incisions with
sloped outside edges pro-
vide an illusion of bas-relief, but in fact, the figure
does not rise above the
surface. 91 cm x 71 cm x
58 cm. Abri Labattut
(Dordogne), France.
Gravettian. American
Museum of Natural His-
tory. Photo by R. White.
(17)

Fig. 139 Engraved limestone slab with a human figure whose sex is indeterminable on the right and an apparently pregnant woman on the left. 22 cm long. Terme-Pialat (Dordogne), France. Gravettian or possibly Aurignacian. Musée du Périgord. Photo by A. Roussot. (189)

Fig. 140 Horse, seemingly a pregnant mare, engraved on limestone block, with lines and cup marks above the back and extending across the neck. Three sides and the back of this block are also deeply engraved. About 25 cm long. Terme-Pialat (Dordogne), France. Gravettian or possibly Aurignacian. Royal Ontario Museum. Photo by R. White. Drawing by L. Quagliarello. (209)

Fig. 141a,b   The Venus of Lespugue, one of the most elaborate Upper Paleolithic female statuettes, sculpted from mammoth ivory. Facial features are absent and hair (covering much of the face) is indicated by lines extending to the shoulder at the back. The breasts on which both arms originally rested are exaggerated. The thighs are separated from the buttocks by incisions, are covered in back with a garment composed of plaits or braids represented by incised lines. The lower legs are very short and end without clear representation of feet. 14.7 cm long. Lespugue (Haute-Garonne), France. Gravettian. Musée de l'Homme. Photo by J. Oster, MH. (147)

Fig. 142 Yellow, semi-transparent steatite female statuette, without lower legs, arms (the swelling under the breasts may represent hands folded across the abdomen), or facial details. A large tuft of hair falls over the back of the neck. 4.7 cm high. Barma Grande Cave (Liguria), Italy. Gravettian. Musée des Antiquités Nationales. Photo by MAN. (72)

Fig. 143   Figurine of apparently pregnant woman sculpted from dark green steatite, known as "Femme au cou perforé" (Woman with Pierced Neck). The reverse side shows a non-pregnant female. 6.9 cm high. Barma Grande Cave, near Grimaldi (Liguria), Italy. Gravettian. Photo by Peabody Museum, Harvard University. (194)

Fig. 144 "Venus of Tursac," sculpted from a flat amber-colored calcite pebble. Without head, arms, or breasts, but with conical trunk, carefully done hips, a large, low abdomen (suggestive of a late stage of pregnancy), massive buttocks and thighs, and short legs without feet. A pedestal engraved with at least two chevrons and inserted between the abdomen and legs and feet seems to have been used to stand the statuette upright in the ground. This is the only French Upper Paleolithic female statuette recovered with full provenience, which permits chronological precision and an understanding of where the statuette lay in relation to other artifacts. 8.1 cm high. Abri du Facteur (Dordogne), France. Gravettian (Périgordian Vc). Musée des Antiquités Nationales. Photo by MAN. Description by H. Delporte. (104)

Fig. 145 "Venus of Sireuil," sculpted from a flat amber-colored calcite pebble, found near the Cazelle (Dordogne) rock shelter and completely without provenience. The head and one hand have been broken off. The figure is characterized by small, youthful breasts; short, folded, arms; large hips; massive thighs; and almost nonexistent lower legs. A vertical hole through the lower limbs suggests use of supporting pedestal to keep the figure upright. 9 cm high. Goulet de Cazelle (Dordogne), France. Probably Gravettian. Musée des Antiquités Nationales. Photo by MAN. (98)

125

Fig. 146   Site of G.
Lalanne's excavations at
Laussel (Dordogne),
France, in the early
1900s. The human fig-
ures shown here were re-
covered from this area,
but (note the shovels)
procedures were too im-
precise to permit close
understanding of their
exact stratigraphic and
spatial location. Photo by
Lalanne, courtesy of A.
Roussot.

Fig. 147   The "Venus à la
corne" (Venus with the
Horn) as she was found,
engraved on limestone
block facing the shelter
wall. Laussel (Dordogne),
France. Gravettian.
Photo by G. Lalanne.
Courtesy of A. Roussot.
(134)

A fourth figure (Fig. 150), 7 inches high, is in a style similar to the others and seems to represent two figures joined at the hips—like the pictures of royalty on playing cards, according to Henri Delporte (1979). It has been thought to represent a sexual encounter, a birth scene, or merely the reuse of the same block for two different figures.

The last major figure from Laussel (Fig. 151), about 20 inches high but partly broken, is often known as the "Chasseur de Laussel" (Hunter of Laussel) and is generally interpreted as a male, mainly due to the absence of apparent female traits.

Other objects were associated with this cluster of human figures, notably several stylized vulvas and phalluses. Two nearby animal engravings (a doe and a horse) are assumed to have been somehow related to the other images. According to Lalanne and Bouyssonie (1941–1946), all the works of art at Laussel were concentrated in an area 12 meters by 6 meters (about 37 feet by 19 feet), which these authors regarded as some sort of sanctuary.

The very term "Venus" implies that the female figures were portrayals of deities. This interpretation has become implicit to the point where the term "fertility goddesses" is also used. It must be kept in mind that evidence does not exist to confirm or refute this interpretation. However, even if the notion of a fertility goddess is accepted as reasonable, Ucko and Rosenfeld (1972) raise an important question about the fertility issue. All known hunters and gatherers today are much more concerned with limiting rather than increasing their population. It is hard to imagine circumstances in which hunting and gathering peoples would purposely seek to increase population density. However, a look at current world population problems strongly suggests that people do not always strive to match their numbers to available resources.

While the fertility view has come to dominate, there are clearly other possibilities. For example, the presence of both male and female figures at Laussel suggests at least the possibility that the cluster of rock carvings reflects a set of ideas about social and economic relationships between men and women.

Fig. 148 The "Venus à la Corne" as she appears today, looking to her right, with hair falling on left shoulder. Facial details and lower extremities were omitted. The figure is characterized by pendulous breasts and broad hips, angular fingers, and a barbed sign on her right hip. With her upraised right hand, she holds what is thought to be an incised bison horn. Strong traces of red paint are visible. Block: 54 cm long x 37 cm wide x 15.5 cm thick. Figure: 42 cm high. Laussel (Dordogne), France. Gravettian. Musée d'Aquitaine. Photo by J. Vertut (134)

Fig. 150   Engraved on limestone slab "Les deux personnages" (The Two People), one clearly female, the other indeterminate, are more schematic than the other Laussel figures but in the same style. The figures appear joined at the hips. The engraving has been interpreted both as a woman giving birth and as a copulation scene. Figures: 20 cm high; Slab: 45 cm maximum dimension. Laussel (Dordogne), France. Gravettian. Musée d'Aquitaine. Photo by J. Vertut. (137)

Fig. 151 "Le chasseur de Laussel" (The Hunter of Laussel), low-relief figure whose head, feet, and arms are barely visible. Generally considered a male archer or spear-thrower, the figure might be that of an adolescent female. 48 cm high. (Dordogne), France. Gravettian. Musée d'Aquitaine. Photo by Arnaud/ Roussot, MA. (136)

**Fig. 152 Bas-relief
sculpture on limestone
block showing a bovid
(probably a musk ox or
bison) with head lowered
as if to charge at the detailed human figure at
the right. It is part of a
large frieze lining the
shelter's back wall. This
is a fine example of Solutrean bas-relief tech-
niques. Bovids threaten-
ing humans is a common
Solutrean and Magdale-
nian art theme. Roc de
Sers (Charente), France.
Solutrean. Musée des
Antiquités Nationales.
Photo by MAN. Descrip-
tion by J.-J. Cleyet-Merle.
(97)**

Solutrean

During the Solutrean—when many of the stone tools can themselves
be considered works of art—a new and dramatic art form with roots
in the deeply incised and occasional bas-reliefs of the Gravettian
comes to dominate. These are large-scale bas-reliefs located in or
immediately adjacent to living areas. Two southwestern France sites
have yielded friezes composed of these bas-reliefs. The most
spectacular is the site of Roc de Sers in the Charente region. A series of
massive sculpted limestone blocks (Fig. 152) lined the back wall of a
rock shelter; more were found on the slope in front of the shelter. The
blocks were decorated with horses, bison, reindeer, mountain goats,
and at least one human figure—all executed in relief that in some

instances exceeded 15 centimeters (6 inches). A similar discovery (Fig. 153) was made at the site of Fourneau du Diable in the Dronne Valley of the Dordogne region. Here, Denis Peyrony found two large sculpted blocks in a Solutrean living site. The more striking block has two fully executed aurochs (wild cattle) done in a very sophisticated way, with the back of the front animal forming the belly of the rear one. A large flint chisel showing strong evidence of abrasion was found adjacent to the sculptures and was probably used to produce this magnificent work, which seems to have decorated the place where people slept, ate, and cooked. In other words, this was not an art isolated from day-to-day existence. Both the Roc de Sers and Fourneau du Diable sculptures date to the late stages of the Solutrean (about 20,000 years ago).

Many other engraved and sculpted blocks from Solutrean sites are known, most of them done on a smaller scale than those described here. In at least some areas during the Solutrean, the practice of deep cave painting seems to have been significant. The cave of Tête du Lion

Fig. 154 Sculpted bison from the Cap-Blanc frieze, found lying at the base of the wall. 61.5 cm long. Cap-Blanc (Dordogne), France. Magdalenian. Musée d'Aquitaine. Photo by MA. (131)

Fig. 155 Large sculpted horse, part of 15-meter (50-foot) long bas-relief frieze on the rock shelter's back wall. Cap-Blanc (Dordogne), France. Magdalenian. Photo by J. Vertut.

Fig. 156 Heavily worn
flint tools from base of
Cap-Blanc sculptured
frieze, some probably
used in sculpting the figures. Musée d'Aquitaine.
Photo by J. Vertut. (122)

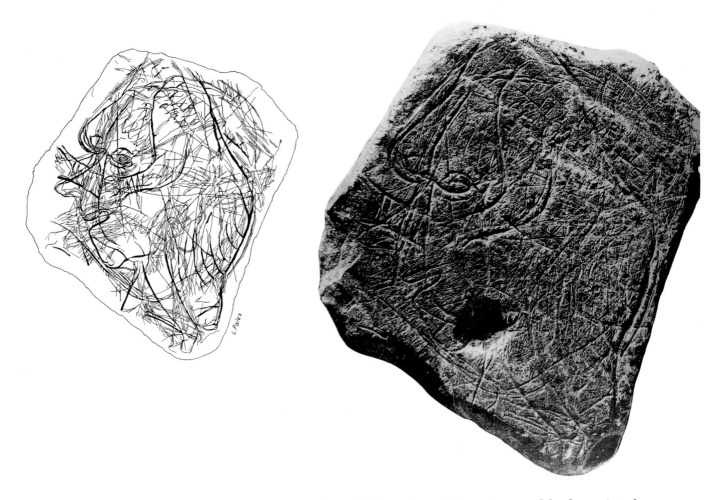

in the Ardèche region of France is one of the few painted caves positively dated to this period. It was discovered during road construction in 1963. Along with a dot and some traces of red paint, it yielded a group of animals painted in red: a deer, an aurochs, and the heads of two ibexes associated with a series of yellow dots. Careful excavations at the base of the paintings by Jean Combier yielded four smears of red pigment and a number of fragments of charcoal, apparently from a torch used by the painters to light their way. The charcoal gave a radiocarbon date of $20,650 \pm 800$ years ago. No stone tools or animal bones were found, indicating clearly that Solutrean people did not live in the cave but merely visited it, probably only once (Combier, 1972).

In the Ukrainian site of Mezin, dated to about 20,000 years ago, there is remarkable evidence for music during this period (Bibikov, 1975). A series of mammoth bones, supposed to have been percussion instruments, were found in a dwelling structure made from

mammoth bones, together with two antler rattles and a rattling bracelet, or castanet. The dwelling, for obvious reasons, is interpreted as having served ceremonial purposes.

Magdalenian

At least 80 percent of all the known Upper Paleolithic art dates to the Magdalenian period, beginning around 18,000 years ago. The Magdalenian exhibits a richness and diversity of art forms; there seems to have been much chronological change in the media employed and the locations chosen for artistic production.

Early phases of the Magdalenian witnessed some of the period's most remarkable deep cave painting, notably that of Lascaux Cave, painted some 17,000 or more years ago. Curiously, the people responsible for the spectacular Lascaux paintings seem not to have produced much in the way of portable art. In the early Magdalenian, there are a few engravings on stone slabs and bone plaquettes, but decoration is restricted for the most part to functional implements, particularly spear-points—and then it is most often schematic or nonanimal in subject matter. To some degree, this can also be said for the cave of Altamira in Spain (Fig. 122), where there is an abundance of engraved objects, almost all carrying abstract designs.

The middle phases of the Magdalenian period, beginning about 15,000 years ago, saw a continuation of deep cave painting. During this period—especially in the French Pyrenees—the deepest ground penetration occurred (Fig. 158). There is also strong evidence for the re-emergence of the large-scale bas-reliefs that were common at the end of the Solutrean. For example, the rock shelter of Cap-Blanc in the Dordogne (Figs. 154–155) is decorated with a frieze of deeply sculpted animals along its entire length of 15 meters (49 feet). The frieze contains several large horses and bison along with several other less easily identified animals. Like the Solutrean bas-reliefs at Fourneau du Diable, this remarkable work seems to have adorned the scene of everyday activities, although it may have some relation to the human burial found at its base.

Fig. 159a Limestone block in form of head with what is believed to be a half-felid/half-human face. The head faces a complex construction partially made up of rosettes composed of cylindrical mounds of packed earth. El Juyo (Santander), Spain. Middle Magdalenian. Photo by L. Freeman, © Institute for Prehistoric Investigations.

Fig. 159b View of cylindrical mounds that form a rosette below El Juyo stone head. The individual cylinders were probably formed by inverting canlike containers of packed earth. El Juyo (Santander), Spain. Photo by L. Freeman, © Institute for Prehistoric Investigations.

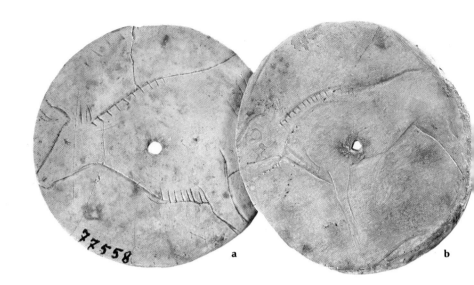

Fig. 160a,b Two sides of bone disk, perforated, with polished rim. On side (a) is an elegantly executed bovid. Horns and ears are rough but in the correct position, as are the eye and nostril, which are represented by dots. Oblique incisions emphasize the back. Side (b), done in the same style but with superior technique, shows a complete calf. Two lines indicate the eye and series of lines indicate back and chest. Both sides together probably represent a cow and her calf. 5 cm diameter. Mas d'Azil (Ariège), France. Late Magdalenian. Musée des Antiquités Nationales. Photo by MAN. Description by H. Delporte. (100)

Provocative new elements of Middle Magdalenian symbolic expression have recently been described by Freeman and Echegaray based upon careful excavations at the site of El Juyo (Figs. 159a,b), near Santander in Spain. In a level dated to 14,000 years ago, they uncovered a series of complex structures that they suggest comprise a "sanctuary." One of these structures included alternating layers of "offerings" (animal bones) and "fill" (rosettelike features made by filling hard-sided cylindrical containers with earth and turning them upside down—as a child might do when building a sand castle on the beach). In the words of the excavators (Freeman and Echegaray, 1981),

Several layers of these rosettes alternate with at least five layers of offerings; the mound built up in this way was 75 centimeters deep when finished. Atop its outermost layer we found several intact bone spear-points and large lumps of red ochre. The whole earth mound was encased in a shell made of clay, reinforced by stone slabs and fragments of long or flat animal bones; the bones used are much larger on average than the usual size of bone fragments in the mound or elsewhere in level 4. Between the dark earth making up the mound and the yellowish clay of its casing, we found about twenty-five whole bone spear-points, placed flat and apparently forming a line around the upper part of the mound.

Overlooking the structures was a stone roughly transformed into what the authors believe to be a half-human, half-feline face.

Middle Magdalenian people also created a rich and spectacular body of smaller portable art objects (Figs. 160, 162–171). Especially

Fig. 161 Engraving of one complete animal, a running lion, between two incomplete felines on a rib split longitudinally. The head is detailed, with whiskers, ears, and tufts of fur. The coat is done with small, aligned markings. An incised geometric figure appears above the lions—something rarely associated with figurative art. Dates to 10,900 B.C. 13.1 cm long. La Vache (Ariège), France. Late Magdalenian. Musée des Antiquités Nationales. Photo by MAN. Description by J.-J. Cleyet-Merle. (109)

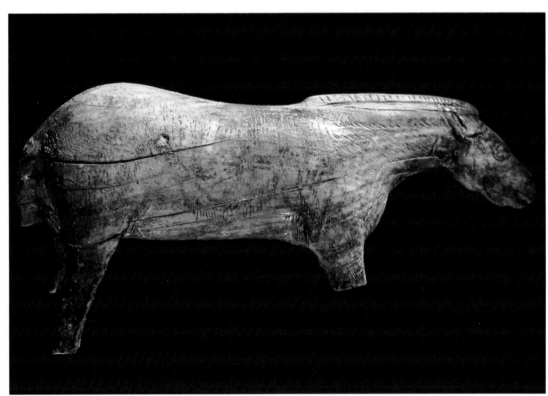

Fig. 162 Horse sculpted in the round in ivory. The neck is horizontal, and the legs, with lower parts broken, are separated from each other. Details of the coat are indicated only on one side. The mane is formed by hatched lines on both sides. The tail is unfinished and roughed in. 7.3 cm long. Lourdes (Hautes-Pyrénées), France. Middle Magdalenian. Musée des Antiquités. Photo by MAN. Description by H. Delporte. (83)

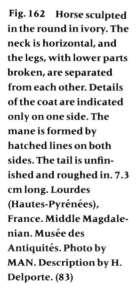

141

Fig. 163 Spear-thrower
with bison licking its
flank, sculpted of rein-
deer antler. This sculp-
ture was constrained by
the antler's shape; the
artist had insufficient
space to sculpt the head
and solved the problem
by having the animal
turn back on its flank and
doing the head in light
relief on the animal's
body—a remarkably so-
phisticated technique.
10.5 cm long. La Made-
leine (Dordogne), France.
Middle Magdalenian.
Musée des Antiquités
Nationales. Photo by
MAN. Description by H.
Delporte. (93)

Fig. 164   Cutaway engraving of horse's head on a flat piece of bone. This piece, only .5 mm thick, is engraved on both sides. The ears are missing. Eye, nostril, mouth, beard, and coat are well portrayed by fine incisions. Narrow incised lines decorated with oblique incisions and running from muzzle to ear are considered by some to be part of a halter. 4.6 cm long. St. Michel-d'Arudy (Basses-Pyrénées), France. Middle Magdalenian. Musée des Antiquités Nationales. Photo by MAN. Description by H. Delporte. (88)

Figs. 165,166 Two cutaway sculptures of horse heads in bone, probably segments of necklaces or other adornments. 5.0 and 7.7 cm long. Enlène (Ariège), France. Middle Magdalenian. Musée de l'Homme. Photo by J. Oster, MH.(186)

Figs. 167a,b Bone fragment engraved on both sides. On side (a) is a bison with massive lowered head, seemingly ready to charge. Eye and nostril are well executed, as are huge horn and heavy mane. Small markings indicate the animal's coat, and two barbed signs appear on its flank. A series of incisions in front of the mouth may indicate the animal's breath. Side (b) shows two human figures—a head and partial torso with arms, and an animallike face, possibly a female, following a nude woman with head missing. On her thigh is a barbed sign identical to that on the bison. 10.5 cm long. Isturitz (Basses-Pyrénées), France. Middle Magdalenian. Musée des Antiquités Nationales. Photo by MAN. Description by H. Delporte. (115)

Figs. 168–171 Antler batons of unknown function sculpted in complex geometric spiral patterns unusual for the Upper Paleolithic. One baton (171) has what may be a very "geometricized" animal head at the top of the decoration. Isturitz (Basses-Pyrénées), France. Middle Magdalenian. Now in unknown private hands. Photo by J. Vertut.

Fig. 172 Engraved horse on limestone slab with legs repeated several times to give the illusion of running or swimming if seen in unsteady light. Block: 33 x 43 cm; Horse: about 16 cm long. Limeuil (Dordogne), France. Late Magdalenian. Logan Museum, Beloit College. Photo by R. White. (64)

Fig. 173 Parts of two horses engraved on bone fragment with special attention to details of the animals' coats. Incisions for head of animal on left extend onto fracture surface, indicating engraving after the bone was broken. 6.4 cm long. Limeuil (Dordogne), France. Late Magdalenian. American Museum of Natural History. Photo by R. White. (21)

146

Fig. 174 Fragment of a limestone block with engraving of the front of an animal, probably a young ibex. Block: about 10 cm high. Limeuil (Dordogne), France. Late Magdalenian. Logan Museum, Beloit College. Photo by R. White. (65)

Fig. 175 Engraved reindeer on limestone slab. Slab: 29 cm long. Limeuil (Dordogne), France. Late Magdalenian. Musée des Antiquités Nationales. Photo by MAN. (89)

Fig. 176 Schematic silhouettes of headless, footless, and often breastless females deeply engraved in limestone block. Roche de Lalinde (Dordogne), France. Late Magdalenian. Field Museum of Natural History. Photo courtesy of Field Museum of Natural History, Chicago. (23)

Fig. 177 Schematic engraving of female, from a site adjacent to Roche de Lalinde. Gare de Couze (Dordogne), France. Late Magdalenian. Musée Nationale de Préhistoire, Les Eyzies, France. Not in exhibition. Photo by Alain Roussot.

149

Fig. 178 Bone fragment engraved with red deer head and detailed and delicate images (lower right) that might be birds or flowers, thus suggesting a late spring/early summer image (Marshack, 1972). 3.8 cm long. Fontarnaud (Gironde), France. Late Magdalenian. Musée d'Aquitaine. Photo by A. Roussot. (130)

Fig. 179 Reindeer with head broken off in front of two branching images, possibly trees or bushes, engraved on thin antler plaque or "spatula" fragment. 7.8 cm long. Rocher de la Peine (Dordogne), France. Late Magdalenian. Logan Museum, Beloit College. Photo by R. White. (62)

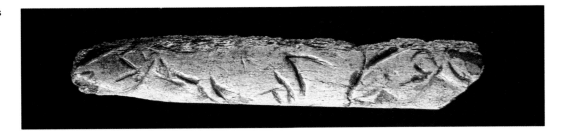

Figs. 180,181 Fragments of engraved antler with highly geometricized animals (180): horses; (181): cervids in sequence. Top: 9 cm long; Bottom: 10 cm long. Laugerie-Basse (Dordogne), France. Middle to Late Magdalenian. Musée de l'Homme. Photo by J. Oster, MH. (169, 175)

151

Fig. 182 Fragment of pierced baton in reindeer antler sculpted in form of human head. 6.5 cm long. Roc de Marcamps (Gironde), France. Magdalenian. Musée d' Aquitaine. Drawing and photo by A. Roussot. (139)

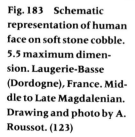

Fig. 183 Schematic representation of human face on soft stone cobble. 5.5 maximum dimension. Laugerie-Basse (Dordogne), France. Middle to Late Magdalenian. Drawing and photo by A. Roussot. (123)

noteworthy are the finely incised limestone slabs and bone and antler
objects sculpted both in bas-relief and in the round. This period
witnessed the re-emergence of many female images—virtually
unknown for the Solutrean and Early Magdalenian. These are usually
engraved, although bas-relief is known. Surprisingly, many of these
images have some of the defining characteristics of their Gravettian
counterparts some 8,000 to 10,000 years earlier. Representations of
adolescents are certainly known, however—for example, from
Laugerie-Basse. Heads and lower limbs are sketchily portrayed, if
portrayed at all. The figures are generally very robust, with large hips
and large, pendulous breasts. These are best known from the French
site of La Marche (Fig. 157), where the painstaking work of Léon Pales
has completely upset the old notion that humans were rarely
represented. He has clearly documented 110 more or less realistic
human representations engraved on limestone slabs (Pales and Tassin
de Saint-Pereuse, 1976).

The spear-throwers from Enlène (Figs. 42,45,46) are from this
period, as are finely sculpted horse-head pendants (Figs. 164–166)
from Labastide, Laugerie-Basse, and other sites. These horse heads—
as well as horses' legs—were usually carved from hyoid (throat) bones
of horses (another example of "iconic congruity") and were most
often perforated for suspension.

The final phases of the Magdalenian, after about 13,000 years
ago, seem to indicate the abandonment of deep caves as centers for
artistic activity. As Leroi-Gourhan (1967) has clearly documented,
most of the Late Magdalenian art (Figs. 178, 179) is found at cave
entrances and in rock shelters and open sites, always in areas
illuminated by daylight. Moreover, this late art—mostly in the form
of engraved slabs and cave walls—takes on what Leroi-Gourhan calls

Fig. 185 Male bear with several dartlike signs on its side, engraved on heavily burned limestone slab. Grotte des Eyzies (Dordogne), France. Late Magdalenian. Slab: about 16 cm long. Logan Museum, Beloit College. Photo by R. White. (37)

hyperrealism. Both the nature and location of this late art are well illustrated by finds from Limeuil (Figs. 173–175) and Teyjat. There is room for caution here, however, since about a third of the slabs from Limeuil were ignored in the publication on the site (Tosello, 1985; White, 1986). These include many representations that are not realistic or even easily decipherable. In contrast to this apparent realism with respect to animal art, the very end of the Magdalenian sees a highly conventionalized— even schematized—treatment of the human and especially the female form, consisting of a very few outlines of generally headless and footless women, as well as some schematic human faces (Figs. 176, 177, 182, 183, 186, 187).

Animal forms at the extreme end of the Magdalenian are also frequently very schematic. A common example is the engraving of animals following each other (Fig. 180, 181). As often as not, they are comprised of only a few very essential lines, a style that contrasts with the detailed treatment of animals' fur and other features earlier in the Magdalenian.

Fig. 186 Headless mammoth-ivory statuette "La Venus impudique" (The Shameless Venus) that contrasts sharply with the many figures representing large and fully adult women in this and earlier periods. 7.7 cm tall. Laugerie-Basse (Dordogne), France. Middle to Late Magdalenian. Musée de l'Homme. Photo by J. Vertut. (173)

Fig. 187 Human figure in vertical position, engraved on long, triangular river cobble. The head shows two profiles, one human and one animal, possibly representing a mask. Two other faces are also engraved. 9.6 cm long. La Madeleine (Dordogne), France. Magdalenian. Musée des Antiquités Nationales. Photo by MAN. (99)

155

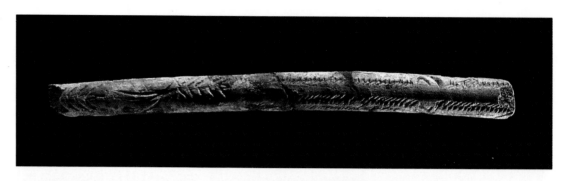

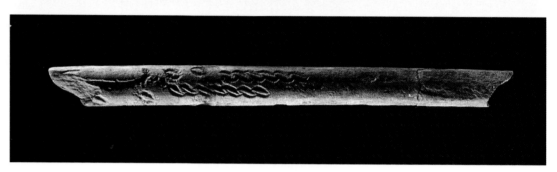

Animal art virtually disappears at the end of the Magdalenian, around 11,000 years ago. The succeeding period, the Azilian, exhibits very little in the way of figurative art apart from small, flat river pebbles (Fig. 193,194) that bear a variety of abstract signs and combinations of such signs. Recent work by Couraud (1985) demonstrates a redundant structure or syntax to the decoration of these objects, suggesting some form of communication or notation system. This research seems to indicate that Azilian pebbles comprised a complex and sophisticated symbol system—far from reflecting artistic degeneration, an argument put forth by many previous scholars.

Recurrent Patterns in the Art: The Search for Structure

The portable art of the Magdalenian shows considerable patterning (Leroi-Gourhan, 1967). Decoration is simple and schematic on object with short lives, such as spear-points, for example. On objects expected to last, decoration is painstaking and realistic.

Different animals were apparently considered appropriate in different contexts. Bison, for example, are often found on stone plaquettes and cave walls but are nearly absent on other media. Horses dominate representations on pierced antler batons and are found on all other media except the numerous semicylindrical rods of split antler. Combinations of motifs often recur on the pierced batons; phallus/fish and horse/reindeer are the most common. The horse/bison combination never exists, even though this is the most common combination on cave walls. Reindeer, the most frequent food animal, are most commonly seen on engraved limestone slabs but are rarely painted. Leroi-Gourhan takes this patterning to reflect a complex body of conceptions operating within the context of a well-developed oral tradition—in simpler terms, a rich body of shared knowledge, belief, myth, and story. It is difficult to disagree. One of Leroi-Gourhan's (1967:83) subtlest recognitions with respect to patterning in Upper Paleolithic art is among his most important:

details in the depiction of the animals' hair and hide recur throughout the range of the representations and they differ for each species. The horses on the

Fig. 190 Ibex painted in red and designed to make use of the stalagmitic ripples in the stone—an example of "iconic congruity." Cougnac Cave (Lot), France. Magdalenian. Photo by J. Vertut.

Fig. 191 Two male bison painted in the "nave" at Lascaux, seemingly running away from each other. Posible a rutting scene. Ground surface is implied by difference in rock color. Composition is nearly 2.5 m long. Early Magdalenian. Photo by J. Vertut.

spear-throwers or on the carved silhouettes, the ibexes and the reindeer, impose the existence during the Middle Magdalenian of a real syntax of drawing. This finding is far removed from the idea one might form of a primitive draftsman who, suddenly inspired by the sight of an appetizing bison, proceeds to draw it with his flint burin. When we examine these horses and bison carefully, it is obvious that they are just as "academic" as the horses in English sporting prints or in Chinese painting.

In examining the art, we often overlook this fact. The realism of the Lascaux paintings is often extoled; in fact, they are highly conventionalized (not to be confused with simplified). These conventions are repetitious and consistent for any given period and indicate a well-defined set of standards, artistic norms, subject matter, and media that had to be met if an artistic representation was to be acceptable and mutually understandable (Fig. 191).

Leroi-Gourhan's suggestion of a syntax of drawing is reinforced in some respects by Margaret Conkey's recent studies of nonanimal engravings (Fig. 188,189) on bone and antler from the Magdalenian of Cantabrian Spain. Conkey (1980) found that a core of 15 design elements (such as chevrons and lunates) was employed throughout Cantabria during the entire Magdalenian.

Conkey also recognized what she calls "iconic congruence" in the art. In essence, the natural contours of a rock wall or a piece of bone or ivory often provided the basis for the artist's choice of the animal to be represented (Fig. 190). This approach applies equally to cave art (where protuberances and stalagmites often bound or help to define the image) and to portable art (where the form of a tool suggests that of the animal represented).

Leroi-Gourhan's best-known search for patterning in the art led him to examine 50 of the painted caves in Franco-Cantabria. A similar but independent search was conducted simultaneously by Annette Laming-Emperaire (1962). The goal in both cases was to determine whether the painted caves were structured with respect to the distribution and location of different animals and signs. Were the different caves organized in similar ways? Leroi-Gourhan and Laming-Emperaire arrived at the same conclusion: the distribution of animals and signs in the caves was not random, and certain animals/signs were consistently associated with other animals/signs. Horses were almost always associated with bison and both were found mainly in the central areas; carnivores and humans were almost always pictured in the far reaches of caves. Not infrequently, they could actually predict, before they arrived at a location within a cave, that a certain image would be there. However, caution is advised here. Recent work by Roussot and others (1984) indicates that the species attribution of difficult-to-identify images may have been influenced by preconceived ideas of which animal *should* appear in a certain location.

Fig. 192 Bison with three drip marks, seen as wounds by supporters of the "hunting magic" explanation of art, engraved on clay cave floor about 1 kilometer (⅝ mile) underground. About 60 cm long. Niaux (Ariège), France. Magdalenian. Photo by J. Vertut.

Although there are certainly disagreements over the specifics and the statistical significance of Leroi-Gourhan's results, most prehistorians now accept his argument that the caves were organized according to some preconceived plan and that the walls were not palimpsests of multiple random painting/engraving events. If more than one painting or engraving event occurred, the later ones followed the same pattern as previous events, often resulting in what appears to us to be "messy" superpositioning of images.

Recently, there has been powerful confirmation of Leroi-Gourhan's and Laming-Emperaire's work. In the context of a broader study of Pyrenean cave art, French archeologist Denis Vialou has analyzed in detail the magnificent painted cave of Niaux (1982). The relationship between such variables as paint color, type of sign, species of animal, and location within the cave is so strong and so predictable that Vialou refers to Niaux as a "symbolic construction."

Social Context, Function, and Meaning in Upper Paleolithic Symbolic Expression

Leroi-Gourhan's primary legacy to cave art studies is scientific rigor and quantification in documenting the paintings and engravings themselves (Ministère de la Culture, 1984). Previously, patterns had been observed and discussed, but most often they were anecdotal rather than founded on a solid body of exhaustive observations. While there are certainly preconceptions embedded in some of Leroi-Gourhan's observations and generalizations, the situation was much worse in the work of previous investigators. The older literature is simply full of false generalizations based on a few well-chosen observations. Most often, these generalizations have to do with the motivation for the art.

For the early researchers in the 1860s who were finding portable art objects in Upper Paleolithic sites, the only motivation required was an esthetic one, coupled with enough leisure time to devote to the arts. For anthropologists, this is the most unsatisfying of all of the explanations for Ice Age art. In no society—including our own—is art motivated purely by esthetic pleasure. Moreover, esthetics are culturally relative and are always embedded with social, political, economic, or religious meaning.

By the turn of the 20th century, when cave paintings were first accepted as authentic, a new realization had emerged—largely because anthropologists were learning more about the context of art in non-Western societies. The notion of totemism or a special, even magical, relationship between animals and humans was developing in ethnology. This concept led Salomon Reinach (1903) to argue that the art of the Upper Paleolithic reflected a set of magical beliefs embodied in the notion of "sympathetic magic." Through the art, Reinach proposed, Upper Paleolithic people sought to increase the numbers of game animals and to ensure hunting success. His view seemed to be supported by the observation that most of the cave art concerned food

animals and that it appeared deep underground in places difficult to reach and therefore not decorated for purely esthetic reasons.

Henri Breuil, who dominated the study of Upper Paleolithic art for nearly 60 years, willingly inherited Reinach's sympathetic-magic framework (Breuil, 1952). Until Breuil's death in 1961, this framework was widely accepted. According to Breuil's version, Upper Paleolithic people had penetrated deep underground to perform increase and hunting rituals. Thus, they exercised ritual control over predators. These ceremonies were one of the contexts in which adolescents were initiated into adulthood. Animals were ritually killed by drawing spears or wounds on their sides (Figs. 185, 192). Certain abstract forms were interpreted as traps and nets drawn to ritually capture the painted animals.

If Leroi-Gourhan succeeded in convincing most of his scientific colleagues that the caves were carefully laid out according to a preconceived plan, he was much less successful in persuading them to accept his ideas as to what the patterning means. Although he has modified his position somewhat, he originally attributed sexual significance to the paired associations of different signs and animals in the caves. He suggested that Upper Paleolithic cosmology was one that divided the world into things that exhibited maleness and things that exhibited femaleness (Fig. 184). Male signs and animals were always associated with female signs and animals in just the kind of binary opposition made famous by the French social anthropologist Claude Lévi-Strauss. There is certainly room for asking whether Leroi-

Gourhan's original male/female cosmology is a direct reflection of modern French language and culture, which breaks the world down according to the principle of gender. But these are the problems we encounter when we try to give meaning to patterns observed in the archeological record. We understand the past through conceptions and interpretive frameworks forged in the present.

Recently John Pfeiffer, a science journalist, has emphasized another way of understanding the deep-cave art, one that may complement Leroi-Gourhan's structuralist approach. Pfeiffer (1983) concentrates on the effect that the cave environment has on the senses and the psyche. Today we visit these painted caves with electric lamps and the security of knowing that many people have been there before us. Pfeiffer suggests that the experience was much different for Upper Paleolithic people, who had only the unsteady and flickering light of burning torches and lamps. Pfeiffer proposes that with this kind of light, the animals will appear to move and breathe. In a situation of sensory deprivation and heightened sensory stimulation at one and the same time, the impact of polychrome bison emerging from the darkness must have been awe-inspiring and dramatic. The

Fig. 195a   A stone cobble with complex engraving of several superimposed animals. The most visible figure is a horse. 12.1 cm long. La Colombière (Ain), France. Late Magdalenian. Musée de Brou. Photo by MB. (143)

unbelievable may have been made believable. Also, we should not ignore modern motivations for cave exploration in imagining the motivations of Upper Paleolithic people. Modern speleology often proceeds in a spirit of machismo that is related to gender roles and individual status. It is reasonable to suppose that conquering the underground (whether done by males or females) 15,000 years ago carried with it a certain amount of prestige.

The fact that much of the art exhibits a profound knowledge of animal development and behavior should also be mentioned. In many instances, the season portrayed can be determined by the condition of an animal's coat or interactions between rutting males of an ungulate species. Future research will undoubtedly provide much insight in this area. Clearly, one of the implications is that the art may have served as a kind of permanent storehouse for environmental knowledge that could be passed down from generation to generation.

Alexander Marshack (1972) has investigated the functions of certain pieces of portable art and proposes that certain incised objects served for tracking time by means of a kind of lunar calendar as early as 32,000 years ago—a hypothesis that has been valuable in stimulating thought among normally conservative archeologists and deserves close study (R. White, 1983). Marshack suggests a calendric function for about 15 of the many worked objects he has examined.

One of Marshack's more provocative contributions concerns images that have been superimposed (Figs. 195a,b), especially animal outlines that have been redrawn several times (Marshack, 1985). Referring to this as reuse and renewal, he views it as periodic "ritual overmarking" indicating that the objects concerned were used in some sort of ritual, perhaps even symbolic, killing. Again, while many professional prehistorians remain doubtful (Lorblanchet, 1973), Marshack has stimulated them to rethink their sometimes staid interpretations.

Understanding the meaning and behavioral context of Upper Paleolithic art represents one of the greatest challenges facing archeologists today. This challenge requires us to cast aside many of our preconceptions and to develop new analytical and interpretive

Fig. 195b The other side stone cobble on facing page. It also has many superimpositions— among them a horse and a rhinoceros. 12.1 cm long. La Colombière (Ain), France. Late Magdalenian. Musée de Brou. Photo by MB. (143)

methods. Above all, we must establish a rigorous data base. This requirement entails years of research just to study a single cave, sometimes even a single panel. An important goal of future research must be to reduce our reliance on ethnographic analogies in interpreting the art. As Leroi-Gourhan (1967:35) pointed out, we will thus avoid having the Upper Paleolithic record speak to us " . . . in the accents of 19th-century Tierra del Fuego or the contemporary Sudan."

In seeking to understand the motives for Upper Paleolithic art, we must also steer away from a tendency toward monocausal explanations, a point emphasized by Ucko and Rosenfeld (1972). After all, these were complicated humans, perfectly capable of integrating a number of cosmological, magical, or functional goals within the same image. In this light, we must keep in mind that, while we subsume all of the Upper Paleolithic imagery under one or two classes (e.g., cave versus portable art), it may well be that a multitude of different types of objects and images related to a variety of artistic and ceremonial contexts are represented in what we have discovered over the past 125 years.

The search for answers continues. It is less a process of discovery from the ground than it used to be. Now it is more a matter of tedious analysis, thought, and interpretation. New discoveries yield new insights, carrying with them the provenience information that a scientific archeology provides. But discovery of a new art object within a controlled excavation answers the questions concerning its creation and use—"when and where?" The difficulty of these questions pales in light of queries related to "how and why?" We find ourselves at the frontier of our ability to understand the past. It is an exciting frontier, but it remains to be mapped.

Bibliography

ApSimon, Arthur. (1980). "The Last Neanderthal in France?" *Nature* 287: 271–272.

Bader, Otto. (1964). "The Oldest Burial?" *The Illustrated London News.* September 9, 731.

Bader, Otto. (1970). "The Boys of Sungir." *The Illustrated London News.* March 7, 24–26.

Bahn, Paul. (1977). "Seasonal Migration in South-west France during the Late Glacial period." *Journal of Archaeological Science* 4: 245–257.

Bahn, Paul. (1982). "Inter-site and Inter-regional Links During the Upper Paleolithic: The Pyrenean Evidence." *Oxford Journal of Archaeology* 1: 247–268.

Begouen, Cmte. H., and Henri Breuil. (1958). *Les Cavernes de Volp. Trois Frères, Tuc d'Audoubert, à Montesquieu-Avantes (Ariège).* Paris: Arts et Métiers Graphiques.

Bibikov, S. (1975). "A Stone Age Orchestra." *UNESCO Courier.* June, 8–15

Binford, Lewis. (1982). "Comment on R. White's Rethinking the Middle/Upper Paleolithic Transition." *Current Anthropology* 23: 177–181.

Binford, Sally. (1968). "A Structural Comparison of Disposal of the Dead in the Mousterian and Upper Paleolithic." *Southwestern Journal of Anthropology* 24: 139–154.

Bordes, François. (1961). "Trace de piquet mousterien à Combe-Grenal (Dordogne)." *L'Anthropologie* 65: 484–486.

Bordes, François. (1968). *The Old Stone Age.* New York: McGraw-Hill.

Bordes, François. (1972). *A Tale of Two Caves.* New York: Harper and Row.

Breuil, Henri. (1952). *Four Hundred Centuries of Cave Art.* Montignac: Centre d'Études et de Documentation Préhistoriques.

Capitan, Louis, and Denis Peyrony. (1928). *La Madeleine: son gisement, son industrie, ses oeuvres d'art.* Paris: Nourry.

Combier, Jean. (1972). "La grotte à peintures de la Tête du Lion à Bidon (Ardèche)." *Etudes Préhistoriques* 3: 1–11.

Conkey, Margaret. (1980). "The Identification of Prehistoric Hunter-gatherer Aggregation Sites: The Case of Altamira."*Current Anthropology* 21: 609–630.

Conkey, Margaret. (1981). "A Century of Paleolithic Cave Art." *Archaeology* 34: 18–28.

Couraud, Claude. (1985). *L'Art Azilien.* Paris: XXth Supplement to *Gallia Préhistoire.*

De Beaune-Roméra, Sophie. (1983). *Les lampes du paléolithique français: définition, typologie et fonctionnement.* IIIe Cycle thesis, Université de Paris I.

De Beaune-Roméra, Sophie. (1984). "Comment s'éclairaient les hommes préhistoriques." *La Recherche* 15: 247–249.

Delluc, Brigitte, and Gilles Delluc. (1978). "Les manifestations graphiques aurignaciennes sur support rocheux des environs des Eyzies (Dordogne)." *Gallia Préhistoire* 21: 213–438.

Delpech, Françoise. (1983). *Les faunes du Paléolithique supérieur dans le sud-ouest de la France.* Bordeaux: CNRS Cahiers du Quaternaire, 6.

Delpech, Françoise, et al. (1983). "Contribution à la lecture des paléoclimats quaternaires d'après les données de la paléontologie en milieu continental. Quelques exemples de flores et de faunes d'Ongules pris dans le Pleistocene supérieur aquitain." *Paléoclimats.* Paris: CNRS. pp. 165–177.

Delpech, Françoise and Jean-Philippe Rigaud. (1974). "Étude de la fragmentation et de la repartition des restes osseux dans un niveau d'habitat paléolithique." *Prémier colloque international sur l'industrie de l'os dans la préhistoire.* Edited by H. Camps-Fabrer. Aix: Université de Provence. pp. 47–55.

Delporte, Henri. (1969). *Chefs d'oeuvre de l'art paléolithique.* Saint Germain-en-Laye: Ministère d'État Affaires Culturels.

Delporte, Henri. (1979). *L'image de la femme dans l'art préhistorique.* Paris: Picard.

De Lumley, Henry. (1969). "A Paleolithic Camp at Nice." *Scientific American* 225: 42–59.

Freeman, Leslie. (1973). "The Significance of Mammalian Faunas from Paleolithic Occupations in Cantabrian Spain." *American Antiquity* 38: 3–44.

Freeman, Leslie. (1980). "Habitation Structures and Burials in the Archaic Aurignacian at Cueva Morín/ Santander, Spain." *L'Aurignacien et le Gravettien (Périgordien) dans leur cadre écologique.* Edited by L. Banesz and J. Koslowski. Nitra, Poland. pp. 77–88.

Freeman, Leslie. (1983). "More on the Mousterian: Flaked Bone from Cueva Morín." *Current Anthropology* 24: 366–377.

Freeman, Leslie, and Jesus Echegaray. (1981). "El Juyo: A 14,000-Year-Old Sanctuary from Northern Spain." *History of Religions* 21: 1–19.

Gaussen, Jean. (1964). *La grotte ornée de Gabillou.* Publications de l'Institut de Préhistoire, Université de Bordeaux, No. 3.

Gaussen, Jean. (1979). *Le Paléolithique supérieur de plein air en Périgord.* Paris: XIVe supplement to *Gallia Préhistoire.*

Gerasimov, I., and A. Velichko. (1982). *Paleogeography of Europe During the Last One Hundred Thousand Years* (in Russian with English abstracts). Moscow: Nauka.

Gladkih, Mikhail, Nenel Kornietz, and Olga Soffer. (1984). "Mammoth-bone Dwellings on the Russian Plain." *Scientific American* 251/5:164–175.

Glory, André. (1959). "Débris de corde paléolithique à la grotte de Lascaux." *Memoires de la Société Préhistorique Française* 5: 135–169.

Hahn, Joachim. (1972). "Aurignacian Signs, Pendants and Art Objects in Central and Eastern Europe." *World Archaeology* 3: 252–256.

Harrold, Francis. (1980). "A Comparative Analysis of Eurasian Paleolithic Burials." *World Archaeology* 12: 195–211.

Heim, Jean-Louis. (1976). *Les hommes fossiles de la Ferrassie, Vol I.* Paris: Archives de l'Institut de Paléontologie Humaine, No. 35.

Issac, Glynn. (1978). "The Food-sharing Behavior of Protohuman Hominids." *Scientific American* 238: 90–106.

Julien, Michelle. (1982). *Les harpons magdaléniens.* Paris: XVIIth supplement to *Gallia Préhistoire.*

Keeley, Lawrence. (1980). *Experimental Determination of Stone Tool Uses: A Microwear Analysis.* Chicago: University of Chicago Press.

Klein, Richard. (1973). *Ice Age Hunters of the Ukraine.* Chicago: University of Chicago Press.

Klima, Bohuslav. (1966). "La station paléolithique de Předmosti près de Prerov (Moravie)." *Investigations archéologiques en Tchecoslovakie.* Edited by J. Filip. Prague.

Lalanne, Gaston, and Jean Bouyssonie. (1941–46). "Le gisement paléolithique de Laussel. Fouilles du Dr. Lalanne." *L'Anthropologie* 50: 1–61.

Laming-Emperaire, Annette. (1962). *La signification de l'art rupestre paléolithique.* Paris.

Lee, Richard. (1979). *The Kung San: Men, Women and Work in a Foraging Society.* Cambridge: Cambridge University Press.

Leroi-Gourhan, André. (1967). *Treasures of Prehistoric Art.* New York: Abrams. Originally published in 1965 as *Préhistoire de l'art occidental.* Paris: Mazenod.

Leroi-Gourhan, Arlette, and Jacques Allain, Eds. (1979). *Lascaux inconnu.* Paris: CNRS.

Leroi-Gourhan, Arlette, Fritz Schweingruber, and Michel Girard. (1979). "Les bois de Lascaux." *Lascaux inconnu.* Edited by Arlette Leroi-Gourhan and J. Allain. Paris: CNRS. pp. 185–188.

Lévèque, François, and Bernard Vandermeersch. (1980). "Les découvertes de restes humains dans un horizon castelperronien de Saint-Césaire (Charente-Maritime)." *Bulletin de la Société Préhistorique Française* 77: 35.

Lorblanchet, Michel. (1973). "La grotte de Saint-Eulalie à Espagnac, Lot." *Gallia Préhistoire* 16: 286.

Marshack, Alexander. (1972). *The Roots of Civilization.* London: Weidenfeld and Nicolson.

Marshack, Alexander. (1981). "Ice Age Art." *Symbols.* Winter. Cambridge, Mass.: Peabody Museum. pp. 4–6.

Marshack, Alexander. (1985). "Theoretical Concepts That Lead to New Analytic Methods, Modes of Inquiry and Classes of Data." *Rock Art Research* 2: 95–111.

Mauss, Marcel. (1904–05). "Essai sur les variations saisonières des sociétés eskimos: étude de morphologie sociale." *Année sociologique* 9: 39–132.

Mauss, Marcel. (1923–24). "Essai sur le don, forme et raison de l'échange dans les sociétés archaiques." *Année sociologique. New Series* 1: 30–186.

Mazière, Guy, and Jean-Paul Raynal. (1983). "La grotte du Loup (Cosnac, Corrèze), nouveau gisement stratifié a Castelperronien et Aurignacien." *Comptes rendus de l'Académie des Sciences de Paris* 296 (Series 2): 1611–1614.

Ministère de la Culture. (1984). *L'art des cavernes. atlas des grottes ornées paléolithiques françaises.* Paris: Imprimerie Nationale.

Movius, Hallam. (1973). "The Abri Pataud Program of the French Upper Paleolithic in Retrospect." *Archaeological Researches in Retrospect.* Edited by G. Willey. Cambridge, Mass.: Winthrop. pp. 87–116.

Otte, Marcel. (1981). *Le Gravettien en Europe centrale.* Brugge: Dissertationes Archaeologicae Gandenses, XX. 2 volumes.

Pales, Léon, and M. Tassin de Saint-Pereuse. (1976). *Les gravures de la Marche II: Les humains.* Gap: Ophrys.

Pfeiffer, John. (1983). *The Creative Explosion.* New York: Harper and Row.

Pigeot, Nicole. (1983). "Un débitage de très grandes lames à Etiolles." *Centre de Recherches Préhistoriques, Cahier 9.* Paris: Institut d'Art et Archéologie. pp. 81–96.

Perlès, Catherine. (1977). *Préhistoire du feu.* Paris: Masson.

Pike Tay, Anne. (1984). *Paleolithic Mortuary Practices Across Eurasia: Towards a Social Analysis.* Unpublished Masters thesis on file, Department of Anthropology, New York University.

Popov, A. (1966). *The Nganasan: The Material Culture of the Tavgi Samoyeds.* The Hague: Mouton.

Reinach, Salomon. (1903). "L'art et la magie. À propos des peintures et des gravures de l'age du renne." *L'Anthropologie* 14: 257–266.

Roussot, Alain. (1973). *Préhistoire en Aquitaine.* Bordeaux: Musée d'Aquitaine.

Roussot, Alain, Robin Frost, and Paulette Daubisse. (1984). "Une nouvelle lecture des gravures enigmatiques de Font-de-Gaume." *Bulletin de la Société Préhistorique Francaise* 81: 188–192.

Salhy, A. (1963). "Nouvelles decouvertes dans la grotte de Gargas." *Bulletin de la Société Préhistorique de l'Ariège* 18: 67–74.

Schild, Romuald. (1976). "The Final Paleolithic Settlements of the European Plain." *Scientific American* 234: 88–99.

Sharp, Lauriston. (1952). "Steel Axes for Stone Age Australians." *Human Problems in Technological Change.* Edited by E. Spicer. New York: Wiley. pp. 69–90.

Simek, Jan. (1984). *A K-means Approach to the Analysis of Spatial Structure in Upper Paleolithic Habitation Sites.* Oxford: British Archaeological Reports.

Simek, Jan. (1986). "A Paleolithic Sculpture from the Abri Labattut in the American Museum of Natural History Collection." *Current Anthropology* (in press).

Skinner, Mark. (1981). "Dental Attrition in Immature Hominids of the Late Pleistocene: Implications for Adult Longevity. (Abstract)." *American Journal of Physical Anthropology* 54: 278—279.

Smith, Fred. (1984). "Fossil Hominids from the Upper Pleistocene of Central Europe and the Origin of Modern Humans." *The Origins of Modern Humans: A World Survey of the Fossil Evidence,* Edited by F. Smith and F. Spencer. New York: Alan R. Liss. pp. 137–210.

Smith, P. (1977). "Selective Pressures and Dental Evolution in Hominids." *American Journal of Physical Anthropology* 47: 453–458.

Soffer, Olga. (1985). *The Upper Paleolithic of the Central Russian Plain.* Orlando: Academic Press.

Stordeur-Yedid, Danielle. (1979). *Les aiguilles à chas au Paléolithique.* Paris: XIIIe supplement to *Gallia Préhistoire.*

Straus, Lawrence. (1983). "From Mousterian to Magdalenian: Cultural Evolution Viewed from Vasco-Cantabrian Spain." *The Mousterian Legacy.* Edited by E. Trinkaus. Oxford: British Archaeological Reports. pp. 73–112.

Stringer, Chris, Jean-Jacques Hublin, and Bernard Vandermeersch. (1984). "The Origin of Anatomically Modern Humans in Western Europe." *The Origins of Modern Humans: A World Survey of the Fossil Evidence.* Edited by F. Smith and F. Spencer. New York: Alan R. Liss. 1–50.

Thompson, D.D., and Erik Trinkaus. (1981). "Age Determination for the Shanider 3 Neanderthal." *Science* 212: 575–577.

Tosello, Gilles. (1985). *Les pierres gravées inédites du M.A.N.* Mémoire de D.E.A. de Préhistoire, Université de Paris I.

Trinkaus, Erik. (1984). "Neanderthal Pubic Morphology and Gestation Length." *Current Anthropology* 25: 509–514.

Trinkaus, Erik, and W.W. Howells. (1979). "The Neanderthals." *Scientific American* 241: 118–133.

Tylor, Edward B. (1871). *Primitive Culture* (2 vols.). London.

Ucko, Peter. (1969). "Ethnography and Archaeological Interpretation of Funerary Remains." *World Archaeology* 1: 262–280.

Ucko, Peter, and Andrée Rosenfeld. (1972). *Paleolithic Cave Art.* New York: McGraw-Hill.

Vialou, Denis. (1982). "Niaux, une construction symbolique magdalénienne exemplaire." *Ars Praehistorica* 1: 19–45.

Vialou, Denis. (1985). "Les images humaines paléolithiques." *Revue Phréatique* 33: 3–16.

Weissner, Polly. (1982). "Risk, Reciprocity and Social Influences on !Kung San Economics." *Politics and History in Band Societies.* Edited by R. Lee and E. Leacock. Cambridge University Press. pp. 61–84.

White, Randall. (1982). "Rethinking the Middle/Upper Paleolithic Transition." *Current Anthropology* 23: 169–192.

White, Randall. (1983). "The Manipulation of Burins in Incision and Notation." *Canadian Journal of Anthropology* 2: 129–135.

White, Randall. (1985). *Upper Paleolithic Land Use in the Périgord: A Topographic Approach to Subsistence and Settlement.* Oxford: British Archaeological Reports.

White, Randall. (1986). "Rediscovering French Ice Age Art." *Nature* 320: 683–684.

Yellen, John. (1977). *Archaeological Approaches to the Present.* New York: Academic Press.

Checklist

Where available, the lending institution's accession number is provided in the checklist that follows.

Jacques Allain, Argenton-sur-Creuse, France

1 Reindeer antler section with sliver removed by groove-and-splinter technique. La Garenne, France. Magdalenian.

2 Bird bone section engraved with reindeer ears. La Garenne, France. Magdalenian.

3 Reindeer antler, damaged by removal of three segments with part of skull still adhering. La Garenne, France. Magdalenian.

4 Pierced baton of reindeer antler with indentations, probably hand grips. La Garenne, France. Magdalenian. EB 32

5 Bone burnisher engraved with eels and fish. La Garenne, France. Magdalenian. GB2 24

6 Decorated pendant or spatula. La Garenne, France. Magdalenian. HIIIB 48

7 Pierced baton of reindeer antler engraved with human image. La Garenne, France. Magdalenian. HOB l

8 Engraved "navette" of reindeer antler, probably a tool handle. La Garenne, France. Magdalenian. L1 C1 44

9 Pierced antler baton in form of phallus. La Garenne, France. Magdalenian. L1 C1 68

American Museum of Natural History, New York, New York, U.S.A.

10 Double-ended scraper of flint coated with white patina. 7.3 cm long. Le Solutré, France. Probably Solutrean. 5-2260

11 Reindeer antler with part of skull adhering and splinter removed. 28.0 cm long. Laugerie-Basse, France. Magdalenian. 5-486

12 Harpoon of reindeer antler, barbed on both edges and engraved with cross-hatched geometric designs. Laugerie-Basse, France. Magdalenian. 75.0-1913

13 Three shouldered flint points. 5.2 cm long, 4.6 cm long, 4.7 cm long. Laugerie-Haute, France. Solutrean. Le Ruth, France. Solutrean. 75.0-318a; 75.0-318b, 75.0-331.

14 Broken spear-point of bone. 10.4 cm long. La Ferrassie, France. Aurignacian. 75.0-344

15 Two flint points. 8.4 cm long, 7.0 cm long. La Ferrassie, France. Mousterian. 75.0-344; 75.0-370.

16 Scraper on end of flint blade with lateral retouch. 7.3 cm long. La Ferrassie, France. Unknown provenience. 75.0-417

17 Large limestone block engraved with horse. 91 x 71 x 58 cm. Labattut, France. Gravettian. 75.0-507 (Fig. 138)

18 Flint blade, laterally retouched. 15.5 cm long. Abri Blanchard, France. Aurignacian. 75.0-525

19 Curved point of bone. 16.2 cm long. Abri Blanchard, France. Aurignacian. 75.0-557

20 Dihedral burin of flint. 8.5 cm long. Fourneau du Diable, France. Solutrean or Gravettian. 75.0-578

21 Bone fragment engraved in elaborate detail with two horses. 6.4 cm long. Limeuil, France. Magdalenian. 75.0-580 (Fig. 173)

Gerhard Bosinski, Institut für Ur- und Fruhgeschichte, Köln, West Germany

22 Stone plaques engraved with schematic human forms. Gonnersdorf, Germany. Magdalenian.

Field Museum of Natural History, Chicago, Illinois, U.S.A.

23 Two limestone slabs engraved with schematic female forms. Roche de Lalinde, France. Magdalenian. 196389 (Fig. 176)

24 Necklace of perforated shells and animal teeth. La Souquette, France. Aurignacian. 196800

25 Small limestone block with remnants of paint. Abri Blanchard, France. Aurignacian. 197757

26 Ivory pendant incised with lines. Abri Blanchard, France. Aurignacian. 197867

27 Bone whistle. La Souquette, France. Magdalenian. 203351

28 Limestone block engraved with circular or vulvar image. Abri Blanchard, France. Aurignacian.

Jean Gaussen, Neuvic-sur-Isle, France

29 Cobble floor of dwelling with seven associated tools. About 2 x 2 m. Le Breuil, France. Magdalenian. (Fig. 86)

Institut für Urgeschichte, University of Tubingen, Tubingen, West Germany

30 Feline sculpted of ivory. 9 cm long. Vogelherd, Germany. Aurignacian. (Fig. 114)

31 Woolly mammoth sculpted of ivory. About 5 cm long. Vogelherd, Germany. Aurignacian. (Fig. 115)

32 Horse sculpted of ivory. 4.8 cm long. Vogelherd, Germany. Aurignacian.

André and Arlette Leroi-Gourhan, Collège de France/Musée de l'Homme, Paris, France.
(Kind permission to borrow was granted by André Leroi-Gourhan prior to his recent death.)

33 Rhynconella fossil, grooved for suspension. Grotte du Renne, France. Châtelperronian.

34 Teeth notched for suspension. Grotte du Renne, France. Châtelperronian.

35 Reindeer phalanx (toe bone) carved in form of deer incisor. Grotte du Renne, France. Châtelperronian.

36 Circular pendant carved from bone. Grotte du Renne, France. Aurignacian. (Fig. 109)

Logan Museum of Anthropology, Beloit College, Beloit, Wisconsin, U.S.A.

37 Limestone slab engraved with bear. Block 16 cm long. Grotte des Eyzies, France. Magdalenian. 10495 (Fig. 185)

38 Bird bone with rows of incisions on three sides. 7.1 cm long. Abri Cellier, France. Aurignacian. 10130 (Fig. 132)

39 Long bone splinter with series of paired incisions. 16.1 cm long. Abri Cellier, France. Aurignacian. 10166 (Fig. 129)

40 Pendant, possibly anthropomorphic, made from horse's kneecap. 7.8 cm x 3.8 cm. Abri Cellier, France. Aurignacian. 10175

41 Diamond-shaped, split-based point of bone. 7.0 cm long. Abri Cellier, France. Aurignacian. 10177 (Fig. 36)

42 Large split-based point of bone. 15.4 cm long. Abri Cellier, France. Aurignacian. 10177 (Fig. 33)

43 Split-based spear-point of bone. 15.8 cm long. Abri Cellier, France. Aurignacian. 10177 (Fig. 34)

44 Split-based point of bone engraved with diagonal line. 10.3 cm long. Abri Cellier, France. Aurignacian. 10177 (Fig. 35)

45 Anthropomorphic figure carved of ivory with paired incisions on each side. 5.1 cm long. Abri Cellier, France. Aurignacian. 10180 (Fig. 110)

46 Bird bone with rows of incisions on three sides. 6.3 cm long. Abri Cellier, France. Aurignacian. 10181 (Fig. 133)

47 Split-based point of bone. 13.6 cm long. Abri Cellier, France. Aurignacian. 10336

48 Fourteen pebbles variously painted with red spots and bands. 3.2–9.5 cm long. Mas d'Azil, France. Azilian. 10494 (Figs. 193,194)

49 Flat pebble with numerous scratches and with polished bevel at one end. Mas d'Azil, France. Azilian. 4.7.1

50 Red sandstone slab for grinding red pigment. 10 cm x 7 cm x 2.6 cm. Crozo de Gentillo, France. Magdalenian. 12527 (Fig. 74)

51 Broken point of reindeer antler engraved with abstract linear markings. 15.7 cm long. Crozo de Gentillo, France. Magdalenian. 12535 (Fig. 37)

52 Bone object of unknown function with spatulate tip. 14.2 cm long. Crozo de Gentillo, France. Magdalenian. 12535 (Fig. 69)

53 Point of reindeer antler, its base formed by two converging bevels, with rows of punctuations. 14.0 cm long. Crozo de Gentillo, France. Magdalenian. 12535 (Fig. 38)

54 Bipointed object of unknown function, of bone or reindeer antler. 9.8 cm long. Crozo de Gentillo, France. Magdalenian. 12538

55 Bone needle with eye. 3.9 cm long. Jouclas, France. Solutrean (?). 12560 (Fig. 89)

56 Small bone needle with eye. 2.9 cm long. Jouclas, France. Solutrean (?). 12560 (Fig. 90)

57 Marine shells and a shark's tooth. Various sites in the Lot, France. Solutrean, Magdalenian. 12563; 12599 (Fig. 104)

58 Polished spear-point of reindeer antler. 14.9 cm long. Jouclas, France. Solutrean (?). 12564 (Fig. 40)

59 Two large curved points of reindeer antler. 27.0 and 33.7 cm long. Jouclas, France. Solutrean (?). 12564

60 Flat blade of bone with incisions at both sides of base. 13.3 cm long. Jouclas, France. Solutrean (?). 12574 (Fig. 70)

61 Bone spear-point, its base formed by two converging bevels, engraved with a series of curved lines. 9.7 cm long. Rivière de Tulle, France. Magdalenian. 12609 (Fig. 39)

62 Spatulate fragment of reindeer antler engraved with reindeer and two abstract designs. 7.8 cm long. Rocher de la Peine, France. Magdalenian. 12650 (Fig. 179)

63 One hundred thirty-seven bone, stone, and ivory beads and pendants reconstructed as necklace. Abri Blanchard, France. Aurignacian. 4.5.407 (Fig. 108)

64 Limestone block engraved with multilegged horse and other animal images. 33.0 x 43 cm. Limeuil, France. Magdalenian. 4.7.206 (Fig. 172)

65 Limestone block engraved with an ibex. Limeuil, France. Magdalenian. 4.7.245 (Fig. 174)

66 Three small pieces of hematite, one showing signs of use as a crayon. 1.9, 2.4, and 3.3 cm long. La Madeleine, France. Magdalenian. 4.7.2507 (Fig. 75)

67 Necklace of bear and lion teeth and marine shells. Rocher de la Peine, France. Magdalenian. 4.7.253 (Fig. 107)

68 Tiny bone implement barbed on both ends. 3.9 cm long. Rocher de la Peine, France. Magdalenian. 4.7.274 (Fig. 53)

69 Tiny barbed point of bird bone. 4.8 cm long. Rocher de la Peine, France. Magdalenian. 4.7.276 (Fig. 49)

Musée des Antiquités Nationales, Saint-Germain-en-Laye, France

70 Bone plaque with punctuations and marginal incisions. 10.1 x 2.9 x .6. Abri Lartet, France. Aurignacian. 15 206 (Fig. 131a,b)

71 Pierced baton of reindeer antler engraved with bear and bird images. 14.0 x 2.3 x 1.7 cm. Massat, France. Magdalenian. 35 82 (Fig. 57)

72 Female figurine, sculpted of yellow stone. 4.7 x 1.2 cm. Grimaldi, Italy. Gravettian. 35 308 (Fig. 142)

73 Limestone slab engraved with reindeer. 12.1 x 8.0 x 1.7 cm. Saint Marcel, France. Magdalenian. 46 669 (Fig. 9)

74 Bone pendant engraved with circular design. 5.8 x 1.6 x .3 cm. Saint Marcel, France. Magdalenian. 46 679 (Fig. 111)

75 Sandstone lamp engraved with ibex on undersurface. 17.2 x 11.9 x 4.2 cm. La Mouthe, France. Magdalenian. 50 295 (Fig. 71a,b)

76 Bone spatula in form of a fish. 19.5 x 2.0 x .6 cm. Grotte Rey, France. Magdalenian. 50 296

77 Limestone slab engraved with reindeer. 43.0 x 30.0 x 6.5 cm. Limeuil, France. Magdalenian. 52 829

78 Animal sculpted in reindeer antler. 8.0 x 3.1 x 1.4 cm. Laugerie-Basse, France. Magdalenian. 53 761 (Fig. 102)

79 Pendant sculpted from reindeer antler in form of salmon. 5.6 x 1.3 x .7 cm. Laugerie-Basse. France. Magdalenian. 53 763 (Fig. 103)

80 Pierced baton fragment of reindeer antler engraved/sculpted with bison, back to back. 12.8 x 7.4 x 3.1 cm. Laugerie-Basse, France. Magdalenian. 53 765 (Fig. 58)

81 Pendant of bone engraved with ovals. 7.3 x 1.9 x .4 cm. Laugerie- Basse, France. Magdalenian. 53 774 (Fig. 118)

82 Pierced baton of reindeer antler, with sculpted animal head. 14.4 x 4.4 x 1.6 cm. Le Placard, France. Solutrean. 55 061 (Fig. 64a,b)

83 Horse sculpted of mammoth ivory. 7.2 x 3.5 x 1.8 cm. Lourdes, France. Magdalenian. 55 351 (Fig. 162)

84 Pierced baton of reindeer antler. About 15.0 x 8.0 cm. Abri Blanchard, France. Aurignacian. 56 339 (Fig. 63)

85 Ivory plaque with series of punctuations in meandering pattern. 9.7 x 3.0 x .6 cm. Abri Blanchard, France. Aurignacian. 56 344 (Fig. 130)

86 Spear thrower of reindeer antler carved with headless animal. 10.7 x 7.1 x 1.2 cm. Arudy, France. Magdalenian. 56 384 (Fig. 44)

87 Baton of reindeer antler engraved with horse. 16.3 x 2.7 x 1.7 cm. Arudy, France. Magdalenian. 56 389 (Fig. 67)

88 Flat bone engraved with horse head showing detailed facial decoration. 4.5 x 2.8 x .1 cm. Arudy, France. Magdalenian. 56 405 (Fig. 164)

89 Limestone slab engraved with reindeer. 29.0 x 10.5 x 5.0 cm. Limeuil, France. Magdalenian. 56 751.22 (Fig. 175)

90 Limestone slab engraved with ibex. About 17.0 x 13.0 x 4.0 cm. Limeuil, France. Magdalenian. 56 751.26

91 Limestone block engraved with vulva. 51 cm long. Abri Blanchard, France. Aurignacian. 56 787A (Fig. 127)

92 Limestone block engraved with vulva. 50 cm long. Abri Blanchard, France. Aurignacian. 56 787B (Fig. 128)

93 Spear thrower of reindeer antler carved and engraved with bison licking itself. 10.5 x 7.1 x 2.4 cm. La Madeleine, France. Magdalenian. 56 873 (Fig. 163)

94 Rod of reindeer antler engraved with bear and vulva. 13.2 x 1.6 x .9 cm. La Madeleine, France. Magdalenian. 56 879 (Fig. 184)

95 Flute of bird bone with two holes on one side and four on the other. About 10 cm long. Abri Blanchard. France. Aurignacian. 56 3 (Fig. 136)

96 Large pendant, often interpreted as a churinga, engraved with geometric design. 16.0 x 3.4 x .6 cm. La Roche de Lalinde, France. Magdalenian. 74 482 (Fig. 117)

97 Limestone block with sculpted bison, human and horses. 153.2 x 57.0 x 27.0 cm. Le Roc de Sers, France. Solutrean. 75 042F (Fig. 152)

98 Female figurine sculpted of stone. 9.0 cm high. Goulet de Cazelle, France. Gravettian. 75 664 (Fig. 145)

99 Quartzite river cobble engraved with human-like figure. 9.6 x 3.3 cm. La Madeleine, France. Magdalenian. 76 950 (Fig. 187)

100 Bone disk engraved with cow on one side and calf on the other. 5 cm in diameter x .2 cm thick. Mas d'Azil, France. Magdalenian. 77 538 (Fig. 160a,b)

101 Limestone block with sculpted bison. 79.0 x 54.0 x 20.0 cm. Le Roc de Sers, France. Solutrean. 80 209

102 Pierced antler baton engraved with horses. 31.1 x 6.2 x 3.4 cm. La Madeleine, France. Magdalenian. 81 62 (Fig. 55)

103 Pierced antler baton engraved with snake, horse, human, and abstract designs. 15.7 x 3.8 x 1.8 cm. La Madeleine, Magdalenian. 81 63 (Fig. 56)

104 Female figurine sculpted of calcite. 8.1 x 4.0 x 2.4 cm. Abri Facteur, France. Gravettian. 81 693 (Fig. 144)

105 Spear thrower of reindeer antler sculpted in form of leaping horse. 20.7 x 2.7 x 1.5 cm. Bruniquel, France. Magdalenian. 82 722 (Fig. 43)

106 Pierced baton of reindeer antler. 23.0 x 8.1 x 2.4 cm. La Ferrassie, France. Aurignacian. 83 132 (Fig. 62)

107 Female figurine sculpted of stone, with detailed vulva. 5.6 x 2.0 x 1.4 cm. Monpazier, France. Gravettian. 83 303

108 Sculpted bone baton with five animals in relief, known as "the scepter." 25.0 x 2.0 x 2.2 cm. La Vache, France. Magdalenian. 83 346 (Fig. 41)

109 Fragment of rib engraved with one complete and two partial lions or panthers. 13.0 x 3.4 x .5 cm. La Vache, France. Magdalenian. 83 347 (Fig. 161)

110 Pierced antler baton sculpted with three humans and an aurochs in relief. 30.2 x 4.3 x 2.3 cm. La Vache, France. Magdalenian. 83 364 (Fig. 60)

111 Sandstone lamp engraved with abstract design. 22.3 x 10.7 x 3.3 cm. Lascaux, France. Magdalenian. 83 461 (Fig. 72)

112 Limestone block with negative hand-print outlined in black paint. 30.0 cm long. Labattut, France. Gravettian. 84 625

113 Reindeer antler baton fragment engraved with bison in profile. 15.5 x 3.3 x 2.9 cm. Isturitz, France. Magdalenian. 84 677 (Fig. 66)

114 Horse head sculpted of soft sandstone showing many facial details. 7.8 x 7.2 x 2.3 cm. Isturitz, France. Magdalenian. 84 710

115 Rod of reindeer antler engraved with bison and humans. 10.5 cm long. Isturitz, France. Magdalenian. 84 772 (Fig. 167a,b)

116 Sitting bear sculpted of stone. 5.8 x 2.8 x 2.3 cm. Isturitz, France. Magdalenian. 84 835 (Fig. 112)

117 Bone spear-point engraved with horse on each side. 17.5 x 2.9 x 1.6 cm. Isturitz, France. Gravettian. 84 860 (Fig. 134a,b)

Moravske Muzeum, Brno, Czechoslovakia

118 Female figure sculpted of ivory, known as the "baton a seins." 8.5 cm long. Dolni Vestonice, Czech. Gravettian.

119 Human face sculpted of ivory. 4.6 cm high. Dolni Vestonice, Czech. Gravettian.

120 Head of feline modeled in clay and fired in a hearth. About 6 cm. Dolni Vestonice, Czech., Gravettian.

121 Head of rhinoceros modeled in clay and fired in a hearth. About 4.5 cm. Dolni Vestonice, Czech., Gravettian.

Musée d'Aquitaine, Bordeaux, France

122 Stone "picks" found at base of sculpted frieze. Probably used to create bas-relief. Various dimensions. Cap-Blanc, Magdalenian. (Fig. 156)

123 Pebble of soft stone engraved with human head. 5.5 x 4.8 x 3.8 cm. Laugerie-Basse, France. Magdalenian. (Fig. 183)

124 Two shoulder blades, both covered with red ocher. 21.5 x 17.0 x 2.5 cm. 26.5 x 12.5 x 2.5 cm. Pair Non Pair, France. Gravettian. Daleau Collection.

125 Flute of bird bone. 12.7 cm long. Pair Non Pair, France. Gravettian. Daleau Collection. (Fig. 135)

126 Reindeer antler fragment carved with human head. 6.3 x. 3.8 x 1.6 cm. Roc de Marcamps, France. Magdalenian. Ferrier Collection. 81.33

127 Reindeer skull with embedded bone spear-point. 14.0 x 13.0 x 9.0 cm. Reignac, France. Magdalenian. Hulin Collection. 67.4

128 Lamp sculpted of limestone. 12.3 x 8.3 cm. Grand Moulin, France. Magdalenian. Labrie Collection. 14 723 (Fig. 73)

129 Limestone lamp with depression for fuel. 14.0 x 11.0 x 6.0 cm. Fontarnaud, France. Magdalenian. Labrie Collection.

130 Bone fragment engraved with deer head. 3.8 x 1.9 x .2 cm. Fontarnaud, France. Magdalenian. Labrie Collection. (Fig. 178)

131 Bison sculpted of limestone fallen from frieze above it. 61.5 x 40.0 x 22.0 cm. Cap-Blanc, France. Magdalenian. Lalanne Collection. 61.3 (Fig. 154)

132 Horse head sculpted of limestone fallen from frieze above it. 42.0 x 19.0 x 9.0 cm. Cap-Blanc, France. Magdalenian. Lalanne Collection. 61.3

133 Reindeer antler engraved with wolverine. 21.5 x 10.5 x 7.5 cm. Laugerie-Haute, France. Magdalenian. Lalanne Collection. 61.3

134 Limestone block sculpted in bas-relief with female figure known as "Femme à la corne." 54.0 x 37.0 x 15.5 cm. Laussel, France. Gravettian. Lalanne Collection. 61.3.1 (Figs. 147, 148)

135 Limestone block engraved with female figure known as "Venus à tête quadrillée." 38.0 x 37.5 x 10.0 cm. Laussel, France. Gravettian. Lalanne Collection. 61.3.2 (Fig. 149)

136 Limestone slab engraved with figure known as "Chasseur de Laussel." 47.0 x 25.0 x 8.0 cm. Laussel, France. Gravettian. Lalanne Collection. 61.3.3 (Fig. 151)

137 Limestone slab engraved with figures known as "les deux personnages." 45.0 x 31.0 x 6.0 cm. Laussel, France. Gravettian. Lalanne Collection. 61.3.4 (Fig. 150)

138 Limestone plaque engraved with vulvas. 41.5 x 34.0 x 8.0 cm. Laussel, France. Gravettian. Lalanne Collection. 61.3.8

139 Pierced baton of reindeer antler sculpted with human head. 6.5 x 3.5 x 2.0 cm. Roc de Marcamps, France. Magdalenian. Maziaud Collection. 70.19 (Fig. 182)

140 Reindeer antler section engraved with human head. 3.6 x 2.7 x 2.4 cm. Roc de Marcamps, France. Magdalenian. Maziaud Collection. 70.19

141 Two "navettes" of reindeer antler, probably handles for stone tools, engraved with linear markings. 22.0 x 1.8 x 1.8 cm and 19.6 x 1.5 x 1.5 cm. Roc de Marcamps, France. Magdalenian. Maziaud Collection. 70.19 (Fig. 32a,b)

Musée de Brou, Bourg-en-Bresse, France

142 Pierced antler baton with a baying deer. 24.5 x 3.5 x 2 cm. Les Hoteaux, France. Magdalenian. 945.232 (Fig. 65)

143 Cobble engraved with multiple superimposed animals. 12.1 x 8.2 x 3.7 cm. La Colombière, France. Magdalenian. 952-7 (Fig. 195a,b)

144 Bone fragment engraved with nonanimal designs. 6.5 x 2.8 x 1.4 cm. La Colombière, France. Magdalenian. 952-8 (Fig. 125a,b)

Musée de l'Homme, Paris, France

145 Mineral pigments of various colors found in cave of Lascaux. Various dimensions. Lascaux, France. Magdalenian.

146 Ivory pendant found with one of earliest upper Paleolithic burials. 3.2 x 2.5 x .3 cm. Cro-Magnon, France. Aurignacian. (Fig. 93)

147 Female figurine elaborately sculpted of ivory. 14.7 x 5.5 x. 3.0 cm. Lespugue, France. Gravettian. (Fig. 141a,b)

148 Five flint bladelets with preserved traces of adhesive. 2.0–4.0 cm. Lascaux, France. Magdalenian. (Fig. 31)

149 Impression of three-strand rope, transformed to humus and embedded in clay. 8.2 x 2.5 x 1.3 cm. Lascaux, Magdalenian. (Fig. 48)

150 Necklace of shells found with buried skeleton. About 2 cm each. Cro-Magnon, France. Aurignacian. (Fig. 93)

151 Semicylindrical rod fragment of reindeer antler. 17,4 x 1.5 x .6 cm. Laugerie-Basse, France. Magdalenian. 38 189 1234

152 Bone point fragment, engraved. 13.0 x 2.0 x 1.1 cm. Laugerie-Basse, France. Magdalenian. 38 189 1243

153 Bone point fragment engraved with lines. 16.2 x 1.8x 1.1 cm. Laugerie-Basse, France. Magdalenian. 38 189 1249

154 Burnishing tool of reindeer antler engraved with lines and branchlike designs. 7.5 x 1.0 x .2 cm. Laugerie-Basse, France. Magdalenian. 38 189 1251

155 Spatula of bone engraved with geometric design, thought to represent a stylized fish of the salmon family. 19.3 x 2.3 x .2 cm. Laugerie-Basse, France. Magdalenian. 38 189 1253

156 Reindeer antler engraved with horse. 27.5 x 2.7 x 3.5 cm. Laugerie-Basse, France. Magdalenian. 38 189 1262

157 Pierced baton of reindeer antler engraved with geometric design. 23.0 x 7.0 x 2.0 cm. Laugerie-Basse, France. Magdalenian. 38 189 1264 (Fig. 61)

158 River cobble engraved with a female reindeer. 7.0 x 5.5 x. 1.0 cm. Laugerie-Basse, France. Magdalenian. 38 189 1323

159 Pierced baton of reindeer antler engraved with lion. 20.0 x 2.8 x 1.5 cm. Laugerie-Basse, France. Magdalenian. 38 189 1325 (Fig. 16)

160 Rib engraved with horses. 17.0 x 3.3 x .6 cm. Laugerie-Basse, France. Magdalenian. 38 189 1326

161 Bone fragment engraved with geometric design. 10.8 x 2.0 x 2.2 cm. Laugerie-Basse, France. Magdalenian. 38 189 1345

162 Salamander sculpted of reindeer antler. 15.0 x 2.0 x 2.9 cm. Laugerie-Basse, France. Magdalenian. 38 189 1357 (Fig. 20)

163 Rib fragment engraved with two horse heads. 9.5 x 1.8 x .2 cm. Laugerie-Basse, France. Magdalenian. 38 189 1359 (Fig. 96)

164 Semi-cylindrical antler rod engraved with heads of two cervids. 7.4 x 1.4 x .3 cm. Laugerie-Basse, France. Magdalenian. 38 189 1360 (Fig. 95)

165 Fish-shaped spatula of bone engraved with horse. 15.5 x 1.5 x .1 cm. Laugerie-Basse, France. Magdalenian. 38 189 1362 (Fig. 101)

166 Spear thrower(?) on bone fragment engraved with fish. 5.6 x 2.0 x .4 cm. Laugerie-Basse, France. Magdalenian. 38 189 1363 (Fig. 19)

167 Hyoid (throat) bone with cutaway sculpture of horse head. 6.2 x 3.2 x .3 cm. Laugerie-Basse, France. Magdalenian. 38 189 1364

168 Reindeer antler fragment with horse head sculpted in the round. 7.4 x 3.6 x 1.0 cm. Laugerie-Basse, France. Magdalenian. 38 189 1366 (Fig. 94)

169 Reindeer antler fragment engraved with horses following each other. 9.0 x 1.5 x 1.0 cm. Laugerie-Basse, France. Magdalenian. 38 189 1368 (Fig. 180)

170 Pierced baton of reindeer antler engraved with horse heads and fish. 22.5 x 3.2 x 2.2 cm. Laugerie-Basse, France. Magdalenian. 38 189 1369

171 Pierced baton of reindeer antler engraved with reindeer. 27.0 x 8.5 x 2.0 cm. Laugerie-Basse, France. Magdalenian. 38 189 1370 (Fig. 68)

172 Spear thrower (?) of reindeer antler sculpted with bird. 21.0 x 3.0 x 2.7 cm. Laugerie-Basse, France. Magdalenian. 38 189 1371 (Fig. 17)

173 Female figure sculpted of ivory, often known as "Venus Impudique." 7.7 x 1.7 x 1.3 cm. Laugerie-Basse, France. Magdalenian. 38 189 1372 (Fig. 186)

174 Bird sculpted of reindeer antler. 5.7 x 4.2 x 1.5 cm. Laugerie-Basse, France. Magdalenian. 38 189 1718

175 Pierced baton of reindeer antler engraved with does. 10.0 x 2.8 x 1.2 cm. Laugerie-Basse, France. Magdalenian. 38 189 1718 (Fig. 181)

176 Rib engraved with bison heads. 21 x 3 x 1 cm. Laugerie-Basse, France. Magdalenian. 38 189 1720

177 Flat part of reindeer antler engraved with bison heads. 7.0 x 6.6 x .5 cm. Laugerie-Basse, France. Magdalenian. 28 189 1721 (Fig. 100)

178 Rib engraved with bovids. 20.0 x 3.3 x .2 cm. Laugerie-Basse, France. Magdalenian. 38 189 1722 (Fig. 98)

179 Spear thrower: fragment of reindeer antler sculpted with deer head. 14.5 x 1.9 x 1.4 cm. Laugerie-Basse, France. Magdalenian. 38 189 1723 (Fig. 97)

180 Two semicylindrical fragments of reindeer antler engraved with abstract images. 8.5 x 1.0 x .3 cm. 16.0 x 1.5 x .5 cm. Laugerie-Basse, France. Magdalenian. 38 189 1724, 38 189 1725 (Figs. 188, 189)

181 Two semicylindrical sections of reindeer antler. 12.7 x 1.4 x .6 cm, 16.0 x 1.5 x .5 cm. Laugerie-Basse, France. Magdalenian. 54 10 19; 54 10 13

182 Spear thrower of reindeer antler sculpted with headless ibexes embracing. 9.0 x 7.0 x 1.2 cm. Enlène, France. Magdalenian. 55 33 l (Fig. 42)

183 Spear thrower of reindeer antler sculpted with cervid. 8.7 x 4.0 x 1.8 cm. Enlène, France. Magdalenian. 55 33 2 (Fig. 45)

184 Spear thrower of reindeer antler in form of bird. 8.9 x 2.1 x 1.4 cm. Enlène, France. Magdalenian. 55 33 3 (Fig. 46)

185 Bone fragment engraved with a grasshopper and other images. 10.0 x 4.5 x 1.5 cm. Enlène, Magdalenian. 55 33 4 (Fig. 18)

186 Three bone sections, each with cutaway sculpture of horse heads. 7.7 x 2.5 x .2 cm, 5.0 x 2.1 x .3 cm, 6.0 x 1.7 x .2 cm. Enlène, France. Magdalenian. 55 33 5, 55 33 6, 55 33 7 (Figs. 165, 166)

Musee du Périgord, Périgueux, France

187 Two perforated reindeer phalanges, Abri Blanchard, France. Aurignacian.

188 Limestone block engraved with zoomorphic figure. 71.0 x 38.0 x 20.0 cm. Abri Blanchard, France. Aurignacian. 4708

189 Limestone slab engraved with human figures. 22.0 x 10.0 x 8.0 cm. Terme-Pialat, France. Gravettian. 9897 (Fig. 139)

190 Semicylindrical rod of reindeer antler engraved with horses. 22.5 x 1.2 cm. Le Soucy, France. Magdalenian. A.1566

191 Engraved pendant. 4.1 x 1.5 cm. Laugerie-Basse, France. Magdalenian. A.1902

192 Bone pendant engraved with bison and hunters. 8.75 x 3.7 x .4 cm. Raymonden, France. Magdalenian. A.2104 (Fig. 119)

193 Bone engraved with bison and humans. 4.5 x 1.35 cm. Raymonden, France. Magdalenian. F.465

Peabody Museum, Harvard University, Cambridge, Massachusetts, U.S.A.

194 Female figurine sculpted in dark green steatite. 6.9 cm high. Grimaldi, Italy. Gravettian T879 (Fig. 143)

Jean-Philippe Rigaud, Direction des Antiquités Préhistoriques d'Aquitaine, Bordeaux, France.

195 Group of six flint cores and assorted flint flakes and blades making up a raw material "cache." Le Flageolet I, France. Gravettian.

Royal Ontario Museum, Toronto, Canada

196 Unretouched blade of gray/blue flint. 14.6 cm long. Belloy-sur-Somme, France. Magdalenian. 926.45. 53

197 End scraper/burin combination of smoky gray flint. 7.5 cm long. Le Moustier, France. Solutrean. 929.21.406 (Fig. 26)

198 Transverse scraper of flint. 11.4 x 8.5 cm. Combe-Capelle, France. Mousterian. 929.29.107

199 Large unretouched blade of beige flint. 24.0 cm long. Le Soucy, France. Magdalenian. 929.29.134

200 Hand axe of gray flint. 12.3 cm long. Combe-Capelle, France. Mousterian. 929.29.142 (Fig. 5)

201 Convergent side scraper of flint. 8.5 cm long. Combe-Capelle, France. Mousterian. 929.29.146

202 Side scraper of flint with cortex still adhering. 11.9 x 7.0 cm. Combe-Capelle, France. Mousterian. 929.29.147 (Fig. 7)

203 Side scraper of flint. 12.9 x 9.4 cm. Combe-Capelle, France. Mousterian. 929.29.182

204 Very large brown flint core with only one blade removed. 36 x 14 cm. Le Soucy, France. Magdalenian. 929.29.219

205 Lamp or grinding stone of soft limestone with a series of incisions. 14 x 14 x 8 cm. Le Soucy, France. Magdalenian. 929.29.221

206 Pointed blade of flint with Aurignacian retouch on both edges. 11 cm long. Laussel, France. Aurignacian. 929.29.265

207 Leaf-shaped point of brownish gray flint. 12.9 cm long. Peyreloude, France. Solutrean. 929.29.289

208 Flint point of type known as Font-Robert points. 5.7 cm long. Roc Tombé, France. Gravettian. 929.29.313

209 Limestone slab engraved with horse and with deep incisions on reverse side. 25.0 x 21.7 x 8.4 cm. Terme-Pialat, France. Gravettian. 929.29.318 (Fig. 140)

210 Blade of smoky-gray flint in form of anthropomorph, perhaps a pendant. 12 cm long. Le Soucy, France. Magdalenian. 929.29.341 (Fig. 116)

211 End scraper on blade of flint coated with gray patina. 10.5 cm long. Roc Tombé, France. Aurignacian 929.29.343 (Fig. 25a)

212 End scraper on dark gray flint flake, 6.9 cm long. La Madeleine, France. Magdalenian. 929.29.400

213 Perforator on delicate blade of translucent gray/brown flint. 9.1 cm long. Le Soucy, France. Magdalenian. 929.29.420 (Fig. 28)

214 End scraper on unretouched blade of gray flint. 9.2 cm long. Le Soucy, France. Magdalenian. 929.29.446

215 Leaf-shaped point of gray flint. 5.8 cm long. Plateau de la Sellerie, France. Solutrean. 929.29.460 (Fig. 30)

216 Broken leaf-shaped point of honey-colored flint. 9.3 cm long. Combe-Capelle, France. Solutrean. 929.29.462 (Fig. 29)

217 Leaf-shaped point of white patina-coated chalcedony with "soapy" finish. 9.5 cm long. Unknown location, France. Solutrean. 929.29.466

218 Leaf-shaped point of whitish-gray stone. 8.9 cm long. Unknown location, France. Solutrean. 929.29.472

219 Broken pierced baton of reindeer antler engraved with horse head on one side and animal body on reverse. 10.6 cm long. Le Soucy, France. Magdalenian. 929.29.495 (Fig. 59)

220 Perforator or "pick" on blade of black flint. 11 cm long. La Madeleine, France. Magdalenian. 929.29.503 (Fig. 27)

221 Three harpoons of reindeer antler, barbed on both edges. 12.7 cm long, 14.3 cm long, 19.0 cm long. Le Soucy, France. Magdalenian. 929.29.524; 929.29.528; 929.29.531 (Fig. 54)

222 Fire-reddened limestone lamp with depression for fuel. 14 x 12.5 x 7.5 cm. Le Soucy, France. Magdalenian. 929.29.538

223 Flake scraper of beige flint. 8.9 cm long. Combe-Capelle, France. Mousterian. 929.81.99 (Fig. 4)

224 Side scraper on chunky blade of green/gray flint. 11.4 cm long. Combe-Capelle, France. Mousterian. 929.81.107

225 Flake cleaver of beige flint. 6.7 cm long. Combe-Capelle, France. Mousterian. 929.81.109 (Fig. 6)

226 End scraper on blade of brown flint. 9.7 cm long. Unknown period, unknown location, France. 929.81.170 (Fig. 25)

227 End scraper of banded Bergerac flint. 5 cm long. Laugerie-Haute, France. Unknown period. 929.81.185

Beatrice Schmider, Bourg-la-Reine, France

228 Flint nodule slightly retouched to suggest form of female figure. Marsangy, France. Magdalenian.

229 Two reconstructed flint cores. Marsangy, France. Magdalenian. H 14 15, P 18403

Yvette Taborin, Institut d'Art et Archéologie, Paris, France

230 Large reconstructed flint core. Etiolles, France. Magdalenian.

P. Taquet, Institut de Paléontologie, Muséum National d'Histoire Naturelle, Paris, France

231 Ivory fragment engraved with mammoths. 22.7 x 10.0 x 3.0 cm. La Madeleine. France. Magdalenian.

232 Pierced baton of reindeer antler engraved with seal. 37.0 x 6.0 x 2.5 cm. Montgaudier, France. Magdalenian.